# The Price of Cake

# The Price of Cake

## And 99 Other Classic Mathematical Riddles

Clément Deslandes and Guillaume Deslandes

illustrated by Laure Macé de Lépinay

translated by Lorenzo Croissant

foreword by Cédric Villani

The MIT Press
Cambridge, Massachusetts
London, England

The MIT Press would like to thank the anonymous peer reviewers who provided comments on drafts of this book. The generous work of academic experts is essential for establishing the authority and quality of our publications. We acknowledge with gratitude the contributions of these otherwise uncredited readers.
This book was set in LaTeX by the authors. Printed and bound in the United States of America.

Library of Congress Cataloging-in-Publication Data
Names: Deslandes, Clément, author. | Deslandes, Guillaume, author. | Macé de Lépinay, Laure, illustrator. | Villani, Cédric, 1973- writer of preface. | Croissant, Lorenzo, translator.
Title: The price of cake : and 99 other classic mathematical riddles / Clément Deslandes, Guillaume Deslandes ; preface by Cédric Villani ; art by Laure Macé de Lépinay ; translation by Lorenzo Croissant. Other titles: Énigmes mathématiques corrigées du lycée à Normale Sup'.
English Description: Cambridge, Massachusetts : The MIT Press, [2023] | "Énigmes mathématiques corrigées du lycée à Normale Sup'–published by Ellipses– Copyright 2014, Édition Marketing S.A."– title page verso. | Includes bibliographical references and index.
Identifiers: LCCN 2022028133 (print) | LCCN 2022028134 (ebook) | ISBN 9780262545242 (paperback) | ISBN 9780262373739 (epub) | ISBN 9780262373746 (pdf)
Subjects: LCSH: Mathematical recreations.
Classification: LCC QA95 .D4613 2023 (print) | LCC QA95 (ebook) | DDC 793.74–dc23/eng20221007
LC record available at https://lccn.loc.gov/2022028133
LC ebook record available at https://lccn.loc.gov/2022028134

# Contents

# Foreword

Riddles have nourished our history, our literature, our mythology. From Oedipus, to Sherlock Holmes, to Turing, we continue to acclaim those who accomplished their destiny by solving riddles. Real or imaginary, these heroes doubtless embody the human condition, unrelenting in its attempts to decipher the staggering enigma of the world!

In this philosophical quest, scientists have agreed to travel the road together, with order and precision, complete sharing of information, skepticism in a priori reasoning, and control in a posteriori thinking. Their undertaking encountered considerable success, but it has far from exhausted the mystery—each victory over the unknown engenders its own new crop of questions. Thus, the story of science could be rewritten as an endless succession of enigmas, sometimes solved, and other times, waiting to be solved.

Yet, beyond their solution, riddles also constitute a powerful learning experience in and of themselves. A student defending their thesis, brimming with pride for solving the riddle put forth by their supervisor (or a riddle they gave themselves!), will certainly not contradict me here.

And there is no need to wait until one is in a doctoral program to take advantage of the power of riddles. Like many readers, no doubt, I remember distinctly the delicate mental torture that I inflicted upon myself as a child to solve a problem presented by a teacher or a science magazine for youths. One such problem, a geometry exercise that I finally solved with an ingenious and complex method, earned me bitter chagrin when my teacher showed me a much more simple and elegant solution. Another, regarding the convergence of series, had me fretting for months until the solution appeared to me in my bed, like an epiphany, the day before the start of the school year. Such experiences came to enrich my development at least as much as my classes did.

A riddle is never just an exercise, but an exercise presented with a certain dose of decorum and mystery, whose solution demands some dose of imagination and will impress the mind. And a riddle is also simply the opportunity to work your brain, for a minute, for an hour, or for a week.

What frustration, what exasperation, when one is busy looking for the solution: the eyes glazed-over and the dumbfounded face as one's mind is occupied by a variety of contradictory thoughts racing through, or, perhaps, running in circles. "It really sounds familiar ..." "Ah, yes, of course..." "I've already spent an hour on this." "In truth, I don't see it." "And what if..." "No, it is impossible!" "I'm sure there's a mistake in the statement!" "But, it's inconceivable, how can it ..." "Ah it's unbelievable how stupid I am!" "What a twisted mind this guy has!" "Why on earth am I wasting my time on this thing?" "I am SURE there is a mistake in the statement!" Comic monologues occur in one's brain!

Why do we waste our time? Because we feel so proud when we solve the riddle without peeking at the solution. But also because we feel so good, in spite of the frustration, when we are looking for the solution!

Compendia of riddles are a dime a dozen. But the one by the Deslandes brothers (Mycroft and Sherlock of the math world?) is remarkable in more ways than one.

First, this book is impeccably presented: the riddles are sorted by the concepts at play, funny situations, antiquated or offbeat drawing, hints to set you on the right path, and finally, complete solutions.

But the book is also remarkable because of its variety. Some riddles are famous, whereas others are obscure. Some situations are ridiculous, whereas others are natural. There are problems posed by great mathematicians and problems that arise simply from common sense. Some questions are very old, but others are drawn from recent research articles. And some problems are simple, while others will leave even professional mathematicians drawing a blank.

Above all, this book is remarkable because the riddles in it lead to doing good mathematics. The solutions to these riddles aren't simply a way for the authors to show off or entrap the reader, but they serve to form a theoretical base and to develop new techniques and technologies. As the complexity of the solutions in the book progresses, we will see more and more advanced concepts come into play, touching on all fields of mathematics, including logic, combinatorics, analysis, algebra, probability, and geometry.

One could complain that by placing the theory at the end of the book, we are moving backward relative to the classical order in a mathematics course, which goes from theory to applications. And perhaps we are moving in reverse, but one can also respond that it is only a return to the natural order of things, since mathematical concepts have been developed first and foremost to solve problems: to clear up what seemed to have been plucked from a large hat, and to design the techniques that will allow the human mind, evidently never lacking in ingenuity, to solve all kinds of problems that haven't even been thought of yet.

Dr. Cédric Villani
Fields Medalist

# Preface

This book is a compendium of difficult mathematical riddles. The statements are precise and the solutions are logical, no tricks are involved. Only those elements precisely defined in the statement can be used to find the solution. For example, in the riddle of the principal and the lightbulb, it would be out of the question to use the fact that a lightbulb turned off recently would still be warm, since it is not explicitly stated in the problem that the lightbulb heats up. All events unfold in an ideal mathematical world, and the different protagonists (students, pirates, snakes) are all perfect logicians.

The riddles are ordered by the complexity of the reasoning required and not by the required level of mathematical knowledge. Thus, an outstanding high school student might be stuck on a level 1 riddle if it requires mathematical background they do not have.

At the end of this book, there are mathematical reminders that will allow anyone to acquire the knowledge required to bypass any such a roadblock, and solve most of the riddles (there are a few exceptions, such as the riddle of the ant and the car which requires some mastery of differential equations, a topic too vast to cover here). However, complexity can be subjective, and the reader's experience of the difficulty of riddles might somewhat differ from the difficulty implied by the order in which the riddles are listed.

Even if one is well prepared, it is recommended to take some time to look for the solution. To take a small glimpse at the answer is to take the risk of definitively forfeiting the pleasure of finding the solution later. On the other hand, if the search proves fruitless for some time, there is a chapter of hints purposefully designed to set one on the right path without spoiling the resolution.

Many thanks must be extended to Damien Thillou and Nolwenn Bellégo for their careful proofreading, to Cédric Villani for his foreword, and to

Thibaut Deslandes, Paul Le Corguillé, Jacques Bonassi, Rémi Nollet, and Laurent Domingos.

Please do not hesitate to send your remarks by email to enigmes@gmx.com. The authors wish you as much pleasure in exploring this book as they themselves had in writing it.

# Part I

# Statements

# Level 1

## 1 A Lazy Blacksmith

A blacksmith wishes to craft a chain 15 links long out of these five chains containing three links each:

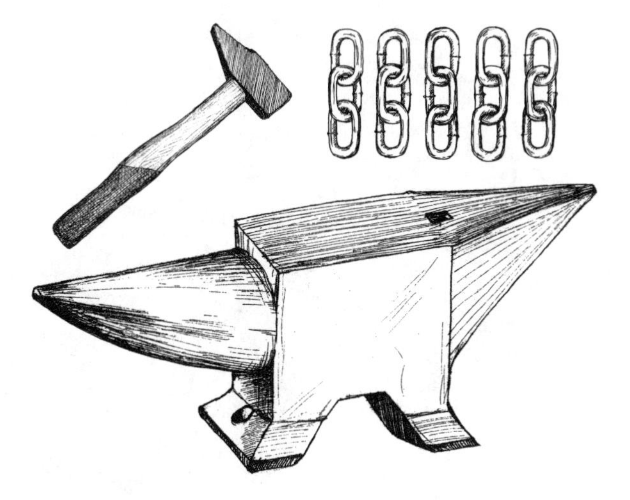

To do this, he can open links and close them again. How should he craft his chain to open as few links as possible?

## 2 Bumper Ants

Fifty ants are dropped at the same time on a stick. Each ant lands on an arbitrary part of the stick and starts moving in an arbitrary direction. When two ants bump into each other, they immediately bounce off each other into opposite directions. The stick is 1 meter long and the ants move at a speed of 1 m/min. When an ant arrives at an end of the stick, it falls off.

What is the longest time it could take for all the ants to fall off?

*For example, if only one ant dropped on the stick instead of fifty, it will take at most 1 minute to fall off. Indeed, the worst case happens when the ant lands on the right end of the stick and starts moving left.*

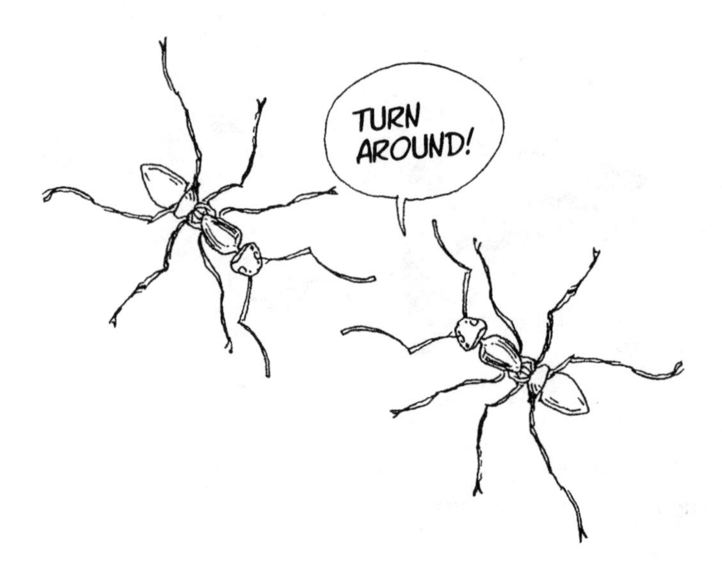

# 3 The Price of Cake

Three friends are preparing for a party. The first friend bakes five cakes, and the second one bakes another three; all the cakes are identical. Since the third friend doesn't know how to bake, she works out that if she chips in \$8 to be split between her friends, they will all have spent the same amount of money.

How must her friends split the \$8 between them so that everyone has contributed the same amount overall?

*She can, for example, give* \$5 *to her friend who made five cakes and* \$3 *to her friend who made three, but is this fair?*

# 4 The Ketchup Fry

A hungry diner is finishing up his plate of French fries. With only one fry left, he decides he needs a bit more ketchup. Being a little clumsy, he accidentally empties the whole bottle onto his plate. The diner notices that the mass of the ketchup makes up 99% of the total mass on his plate, which is 1 kg. After thinking about it for a while, and considering his love of ketchup, he decides to remove just enough ketchup to have 98% of the total mass of his plate be ketchup.

How much ketchup must he remove?

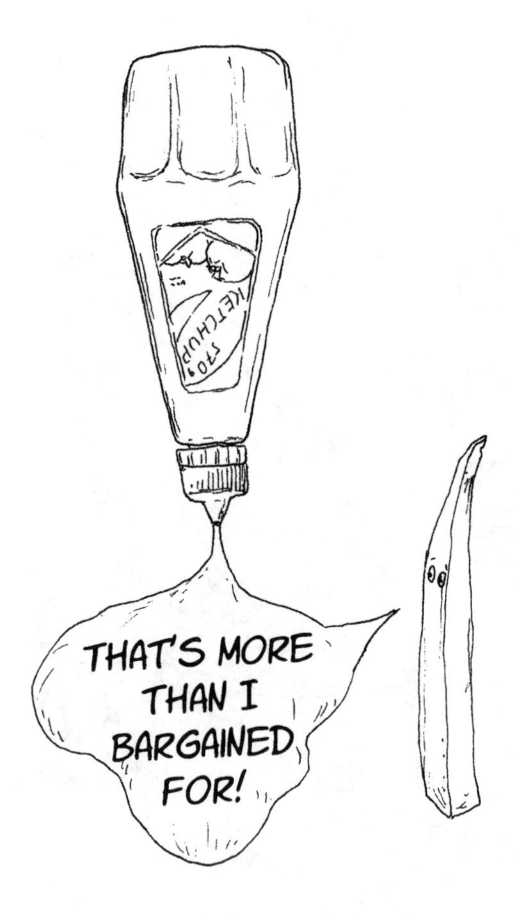

## 5 Going Around in Circles

An explorer walks 1,000 km south, then 1,000 km west, and then 1,000 km north. He then notices that he has come back to his starting point!

What are his possible starting points?

*The North Pole is a possible starting point, but it is not the only one.*

## 6 The Soft-Boiled Egg

A starving adventurer, out in the middle of the jungle, wants to cook herself a soft-boiled egg. She has everything she needs to do so except for a stop-watch! Thankfully, she picked up two vines that she knows from experience take exactly 4 minutes to burn.

How can she time 3 minutes with these two vines?

*The vines take exactly* 4 *minutes to burn entirely, but need not burn at a consistent rate. For example, half of a vine may burn in* 1 *minute, while the other half will burn in* 3.

# 7 A Pill Mix-up

A patient has been prescribed a special course of pills by his doctor. He must take exactly one A pill and one B pill every day for 30 days. One day, he puts one A pill in his hand and then accidentally puts two B pills in the same hand. It is impossible to tell the pills apart; hence, he has no idea which is the A pill and which are the B pills. He only had 30 A pills and 30 B pills to begin with, so he can't afford to throw the three pills away.

How can the patient follow his treatment without losing a pill?

*It is possible to cut pills into several pieces.*

# 8 A Hurried Student

A student bikes to the university every day. He averages 20 km/h on the way there.

How fast must he go on the way back so that his average round-trip speed is 40 km/h ?

# 9 Going Upstream

A Paris *bateau-mouche* follows the same route every day: it goes down a part of the Seine River, turns around, and goes back upstream to its starting point. Today, the current is faster than usual.

Will it take less time, as much time, or more time than usual for the boat to make the round trip?

*It will take the boat less time on the way down but more time on the way back.*

# 10 Boys and Girls

A country decrees that a family can have a second child only if the first is a girl and, in any case, can never have more than two children.

Will this change the ratio of boys to girls?

## 11 *Café au Lait*

A coffee lover has in front of him a cup of coffee and a cup of milk. Both cups contain the same quantity of liquid: 200 mL. He starts by pouring a teaspoon of coffee into the milk and then puts a teaspoon of the resulting milk-coffee mix into the coffee. This teaspoon contains 5 mL.

At the end of this procedure, is there more milk in the coffee or more coffee in the milk?

## 12 *Lait au Café*

The coffee lover has in front of him a cup of coffee and a cup of milk. Both cups contain the same quantity of liquid: 200 mL. This time, he starts by pouring a teaspoon of coffee into the milk and then pours a teaspoon of the resulting milk-coffee mix into the coffee. This teaspoon contains 5 mL.

Is it possible to obtain a cup containing equal amounts of milk and coffee by repeating this procedure as many times as necessary?

# 13 Guardians of the Gates

Imagine two identical doors: behind one is heaven, and behind the other is hell. Each door is guarded by a guardian. One of the guardians always tells the truth, while the other always lies. However, one cannot know which is which.

By asking only one question to only one of the two guardians, how can one determine which door leads to heaven?

# 14 A Kangaroo on a Staircase

A kangaroo is climbing a staircase with 10 steps. He can climb the stairs step by step or jump over a step whenever he wants and as many times as he wants.

How many different ways of climbing the stairs does the kangaroo have?

*For example, there are five different ways to climb a set of stairs with four steps:*

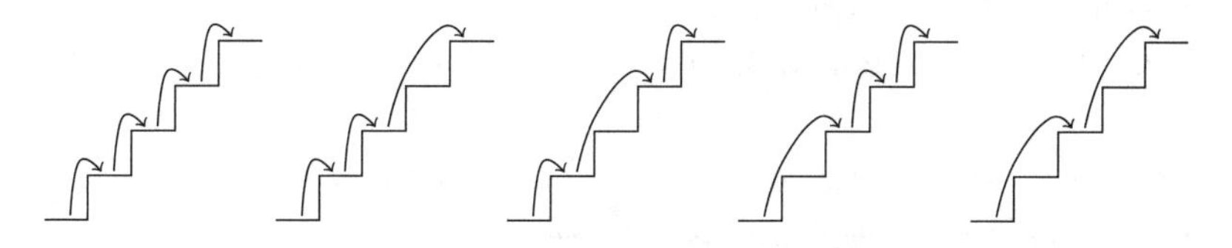

# 15 The Fluttering Fly

Two trains beginning 400 km away from each other are driving straight toward each other, each at a speed of 200 km/h. A fly leaves the first train and flies to the second train at the incredible speed of 400 km/h. When it reaches the second train, it instantly turns around and goes back to the first train, and so on, continuing to fly back and forth until the trains pass each other.

What is the total distance flown by the fly?

# 16 An Hour Early

A little boy finishes school every day at 16:00 (4:00 PM), and his mother comes to pick him up every day at 16:00. One day, the boy's school day ends at 15:00 (3:00 PM). Since he has not let his mother know ahead of time, and knowing she will arrive at 16:00, he decides to walk to meet her instead of waiting for her at school. He walks along his usual route home until he comes across his mother's car. She picks him up, turns around, and drives home. They arrive home 10 minutes earlier than usual.

How long did the boy walk for?

*One will consider that his mother didn't waste any time picking him up and that the car travels at a constant speed.*

# 17 How to Cut Up a Cube

We wish to cut a cube into 27 smaller cubes, like in the following diagram:

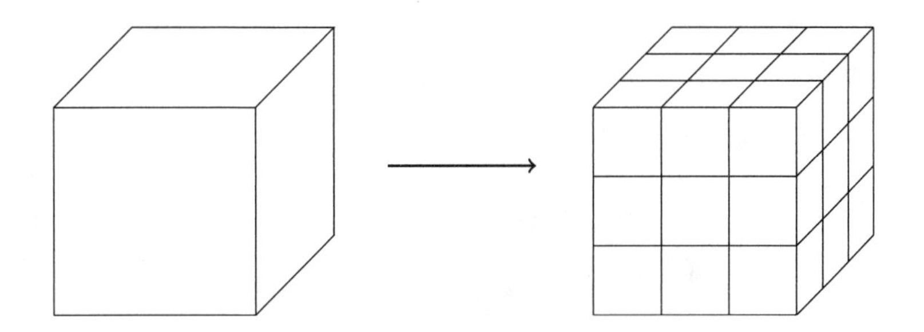

To do so, we have a saw that naturally makes planar cuts: the cuts are perfectly straight. We can cut the cube with six cuts by maintaining the cut pieces in their place, that is, by only separating them once all the cuts have been made.

Is it possible to cut the cube in fewer than six cuts by rearranging the pieces freely between cuts?

# 18 Racing to a Hundred

Two friends are playing a game. Starting from 0, and taking turns, each one adds an integer between 1 and 10 to the current shared total. The first to reach 100 wins.

What strategy can the first player put in place to win every time?

## 19 Fresh Paint!

Three painters must paint a living room together. The first painter would take 2 hours were he to do it alone, the second would take 3 hours, and the third would take 5 hours.

How long will it take them together?

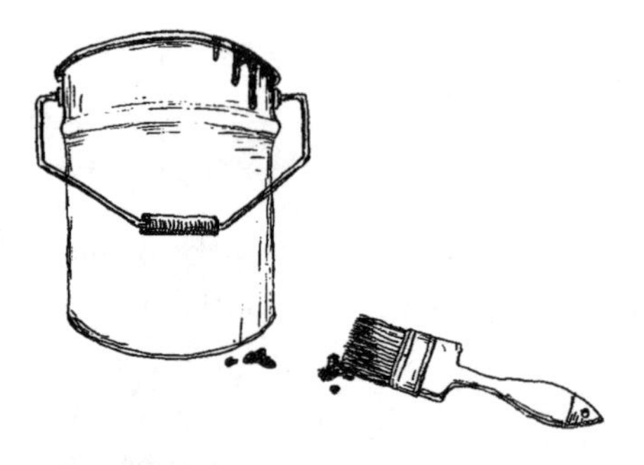

## 20 The Worldly Cocktail Party

During a cocktail party, the guests shake hands as they encounter new people. Each guest shakes each person's hand only once.

Show that, at any instant, there are at least two people who have shaken the same number of hands.

## 21 Same Time and Place

A young executive leaves her house at 8:00 (8:00 AM) to go to work. After a long day of work, she leaves her workplace at 20:00 (8:00 PM) and commutes home via the same route as in the morning.

The executive has an analog watch. Show that on the way back, there is at least one moment where her watch will display the same time as it did in the same place that morning, which means she is in exactly the same place as she was 12 hours earlier.

## 22 Taking It a Coin Too Far

A student is at a job interview. He is offered the following game as a test of his mental acuity. He faces off with his future employer on a square table. Next to them is a box of coins, all round and identical. Each takes turns placing a coin on the table; that coin must not touch any coins already on table. The one who has no room left to place their coin loses. The student goes first.

What strategy can the student adopt to win every time?

## 23 The Birthday Paradox

A classroom has 23 students.

What is the probability that two students have their birthday on the same day of the year?

*For simplicity, we will consider that every year has 365 days. We will also assume that each student has a* $\frac{1}{365}$ *probability of being born on a given day. This problem is in no way a paradox, but it is often called such because the result strongly contradicts intuition. It is worth doing the computation!*

# 24 Fifteen Serpents

Fifteen snakes have encircled a mouse. All the snakes want to eat the mouse, but each snake knows that if it eats the mouse (or anything else), it will fall asleep and might get eaten by one of the other snakes.

What happens?

*Of course, the snakes are starving, but they still prefer to abstain if they will be eaten afterward. We assume that there is always one snake that is faster than the others.*

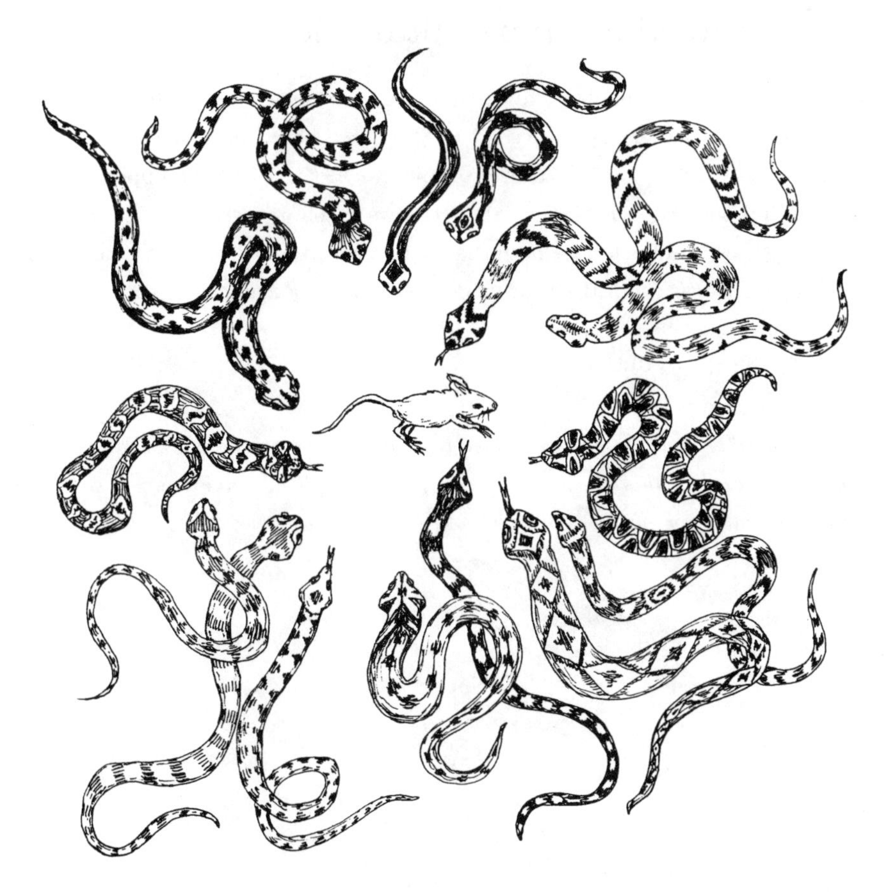

## 25 A Screening Test

An illness affects 0.1% of the population. A pharmaceutical lab offers a screening test which is 99% reliable, meaning it has a 99% chance of returning negative if the person being screened is healthy and a 99% chance of returning positive if the person is sick. The test is always either positive or negative.

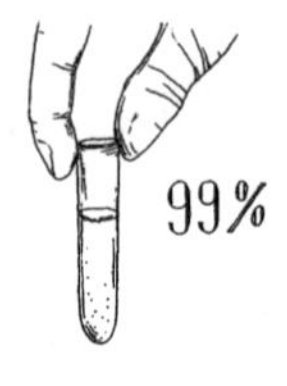


What is the probability that a screened individual who tested positive is indeed sick?

## 26 Shocking Sports Statistics

Two basketball teams play two consecutive matches against each other. In the first game, team A succeeded on a higher percentage of their shots than team B: 93% to 87%. In the second game, they held the upper hand again and succeeded in 73% of their throws to 69% of team B. After the two matches, commentators note that both teams took the same number of shots, but it was team B which made the highest percentage of their shots in general!

How is this possible?

*No need to find the precise values that yield these percentages. The issue is only to explain this phenomenon and find a simple example where it shows up in a blatant manner.*

# Level 2

## 27 Free Throws

A basketball player is practicing free throws. He misses the first throw, but he has an 80% success rate at the end of the day.

Show that, at some point during the day, he had a success rate of exactly 75%.

# 28 The Number Across the Street

In a small village, the postman brings the mail to the schoolteacher's house. She offers him a riddle. "I have three children; the product of their ages is 36, and the sum of their ages is the number of the house across the street. How old are they?"

"I cannot know this," the postman replies, despite knowing the number of the house across the street.

"The eldest plays piano."

"In that case, I know how old each is."

How old are the schoolteacher's children?

# 29 The Unstable Painting

How should one hang a painting with two nails in such a way that removing either nail makes the painting fall to the ground?

*For example, in the figure below, the painting will still stay up even if a nail is removed. This will not do.*

# 30 The Unfaithful Spouses

In an imaginary village, where several couples live, some people are cheating on their spouses! Unfortunately for them, gossip spreads very quickly, and when someone cheats on their spouse everyone in the village knows except, of course, the spouse. But, if the spouse learns of it, the couple separates that very evening, and all the villagers notice the next morning. There were exactly 50 unfaithful spouses in this village living in perfect harmony until one morning someone wrote on a wall in the village: "There is at least one unfaithful spouse in this village."

What happened then?

## 31 A Few Surprising Probabilities

A couple has two children. They are asked if they have at least one daughter, and they respond "yes."

What is the probability that the other child is a boy?

They are then asked if they have a daughter whose name is Sophie. If they answer "yes," what is the probability that the other child is a boy?

Instead of asking them this second question, they are asked simply to give the female first name of one of the children. If they answer Sophie, what is the probability that the other child is a boy?

*As one might suspect given the title of the riddle, the two last probabilities differ! For simplicity, we will assume that all first names have the same probability* $\epsilon$ *of being given at birth.*

# 32 The Monty Hall Problem

During a game show[1] on television, a contestant must choose among three doors presented to her, knowing that behind one of them is a dream car, but behind the other two are worthless token prizes. The candidate makes her choice, and then the host starts his well-known routine: he opens one of the two remaining doors, avoiding the one containing the car so as to not ruin the suspense. The contestant then has the choice to change doors or confirm her first choice.

What must she do?

What happens if now the host doesn't know where the car is and opens a door at random?

Monty Hall

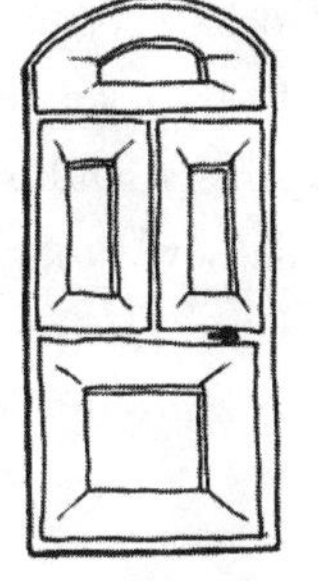

[1]Monty Hall was the host of the show *Let's Make a Deal*, which inspired the problem.

## 33 Square Jigsaws

A jigsaw puzzle manufacturer makes square jigsaws with square pieces. The pieces can be of different sizes, but they are all square and, once assembled, the resulting jigsaw is a square. On each box, the manufacturer writes the number of pieces.

What are the numbers they will never write on a box?

*For example, they will never be able to write the number 2 on their boxes, since it's impossible to make a square out of two smaller squares.*

## 34 The Good, the Bad, and the Ugly

The Good, the Bad, and the Ugly are facing off in a three-way duel. They decide that the Good will shoot first, followed by the Bad, and then the Ugly, and then the Good once more, and so on until there is only one left. When one dies, they are simply skipped. The Good has a 1-in-3 chance of killing his opponent, the Bad has better aim and has a 50/50 chance of killing his opponent, and the Ugly never misses. Each knows the success probabilities of all three protagonists, and only seeks to win the duel. Each may choose his target as he wishes and may even choose to shoot in the air to miss on purpose.

It is up to the Good to start; what should he do?

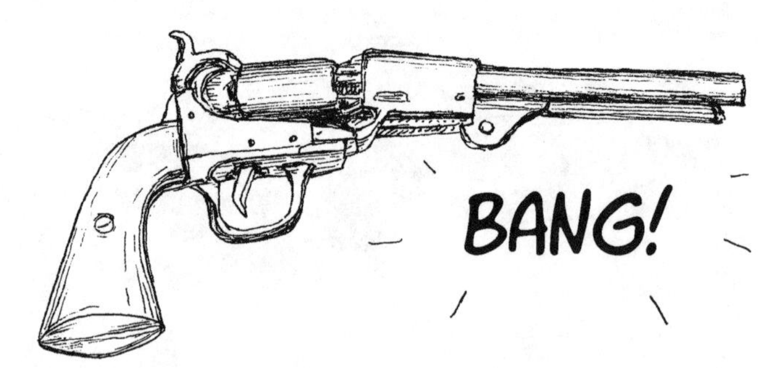

# 35 The Gas Shortage

On the beltway of a deserted city, there are several abandoned gas stations. In total, across all the stations, there is exactly enough gasoline for one car to make one loop of the beltway.

Given a car with an empty tank, and the freedom to pick your starting gas station, is it possible to complete a loop of the beltway?

*Let us suppose that to complete a loop of a* 10 km *long beltway requires* 10 *liters of gas and that there are three gas stations. The first at the zeroth kilometer has* 5 *liters of gasoline, the second one at the sixth kilometer contains* 3 *liters of gasoline, and the last at the eighth kilometer—and thus* 2 km *from the first—contains 2 liters of gasoline. In this case, leaving from the second gas station the loop can be completed.*

# 36 An Impossible Prediction

Two students, who are bored in their probability class, play the following game instead of paying attention. The first begins by mixing a standard 52-card deck and then asks her friend to guess the color (red or black) of the first card. After her friend's prediction, she shows him the first card and asks him again to guess the color of the second card, reveals it, then the third, and so on until the end of the pack.

By adopting the best possible strategy, what is the probability that the student guesses all cards without a single mistake?

## 37 With Three Weighings

Twelve weights are numbered from 1 to 12. All the weights have the same mass except one, which is heavier or lighter than the others. The goal is to discover, with the help of a Roberval balance, which one is different and whether it is heavier or lighter; this must be done in at most three weighings.

How?

## 38 Heads or Tails in the Dark

One hundred coins are placed in complete darkness. They are mixed, but it is certain that 75 are tails and 25 are heads.

How can they be separated into two groups with the same number of heads?

*The two groups formed in this manner need not have the same number of coins.*

## 39 The Principal and the Hats

A school principal, who is also a logician, offers a game to 100 of his students, allowing them, should they succeed, to be freed from detention. The students are placed on a staircase with 100 steps, one student per step. On each student's head, the principal places either a white hat or a black hat. The students do not know the number of black nor the number of white hats handed out. For example, the principal could give out 70 black hats and 30 white hats. Each student can only see the hats of the students on lower steps. Thus the student at the top of the stairs sees the hats of the 99 other students, and the student at the bottom sees none. The principal asks the student at the top for the color of her hat, to which she must answer by saying either "black" or "white," and all the students hear her answer. It is then the student on the next step down who is asked, and so on all the way down. All students who guessed the color of their own hat win. Students may agree on a strategy beforehand.

Find a strategy which can save, with certainty, the maximum number of students.

To go further, consider a variant of the problem with seven possible colors instead of two.

*For example, a possible strategy for students is the following: the first, at the top, says the color of the second's hat, who repeats his color and is freed, the third one says the color of the fourth's hat, and so forth. This strategy frees* 50 *students with certainty but may save more due to pure chance. We count only those saved for sure.*

## 40 The Fort

The contestants of a game show must escape a fortress at sea. In the last test, one contestant faces the game master. Ten coins are lined up on a table before them. Each coin has a value between 1 and 10 that is visible by both players. All the values are present once, but their order is random. The following is an example:

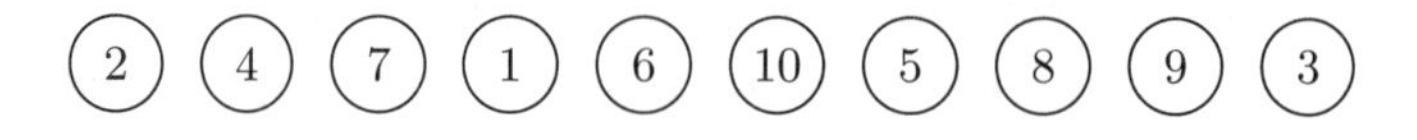


Each player takes turns removing one of the coins situated at an end. The winner is the one who collects the highest total. The contestant starts.

What strategy can he adopt to guarantee that he wins every time?

*In the previous example, he can start by choosing coin 2 or coin* 3*. If he chooses coin* 3*, the game master will have a choice between coin 2 and coin* 9.

## 41 Exact Opposites

Show that there are, at any given time, two places on Earth that are on exact opposite sides of the planet and that have the same temperature.

# 42 Hilbert's Hotel

Let us assume the existence of a very peculiar hotel: it contains an infinite number of rooms! This hotel is made up of an infinite hallway. There is room 1, room 2, room 3, and so on. The hallway never stops, and rooms go on to infinity. Suppose furthermore that this hotel is full! All the rooms are occupied by a client (there are thus an infinite number of clients). A new customer shows up to reception and asks for a room.

How can reception find him a room?

This time, it is not one next customer but infinitely many new customers: there are new customer number 1, new customer number 2, new customer number 3, and so on. How can reception find each of them a room?

*There are infinitely many rooms: the last room does not exist. To find a room for the new customer, we can't build new rooms nor throw out existing customers. The only thing we can do is to reorganize the assignment of rooms.*

## 43 A Chaotic Boarding

An airliner has 100 seats, and its 100 passengers all have a designated seat. Its passengers board one by one. The first passenger, not paying attention to her seat, sits in a seat at random. The second sits in his seat unless it is already occupied by the first passenger, in which case he picks a new seat at random. The third sits in her seat if it is free; if it is not, she chooses a seat at random. This pattern goes on until the last passenger is seated.

What is the probability that the last passenger finds their designated seat free?

## 44 Mutineer Pirates

Ten pirates are sharing a treasure of 100 doubloons. The captain, that is the eldest pirate, decides on the distribution (for example, everything for himself and nothing for the rest). If a strict majority of pirates are opposed to this distribution, a mutiny will break out, and the captain will be thrown overboard. There will then be only nine pirates left, and the second oldest pirate becomes the captain and proposes a distribution. This cycle goes on until a distribution is adopted.

What distribution should the eldest pirate propose?

*All the pirates know the ages of all the other pirates.*

# 45 The Monkey Race

A chimpanzee and a baboon race over a distance of 3.2 km. The chimpanzee runs at the perfectly regular pace of 5 min 45 s per km, while the baboon runs at a potentially irregular pace but traverses each interval of 1 km in exactly 5 min 50 s.

Can the baboon beat the chimpanzee?

*By "each interval," one must understand any possible* 1 km *interval. We know that the baboon will run the first kilometer in* 5 min 50 s, *but also, for example, that he will run the interval* [0.817 km, 1.817 km] *in the same time.*

# 46 Dwarf Thieves

Seven dwarves work in the mine for Snow White. At the end of the day, each dwarf must bring back a gold ingot weighing 100 grams. Some dwarves are honest, but others always steal 1 gram of gold and bring back ingots weighing only 99 grams. In 4 months, Snow White has accumulated approximately 100 ingots per dwarf. She would like to know who steals and who doesn't. In order to do so, she has a magic scale, with a digital display, which is infinitely accurate but only works for one weighing.

How can she determine, in one weighing, who is stealing and who is not?

# 47 Eight Triangles to a Square

Cut a square into eight triangles whose angles are all strictly acute, that is, strictly smaller than 90°.

## 48 How to Tile One's Bathroom

The floor of a bathroom is a square consisting of 64 tiles. Following some work, the tiles of the upper left and lower right corners are removed. The bathroom floor thus takes the following shape:

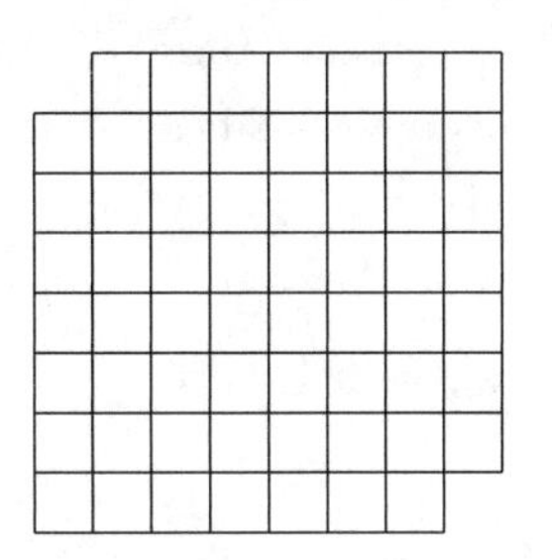

We wish to cover the floor with a domino-shaped tiling:

Is it possible?

## 49 One More Coin

Two brothers are playing heads or tails. The older brother takes 11 coins and only gives 10 to his little brother; thus, the older brother has one coin more. They toss all their coins at the same time and count how many tails they obtained.

Under these conditions, what is the probability that the older brother gets strictly more tails than his little brother does?

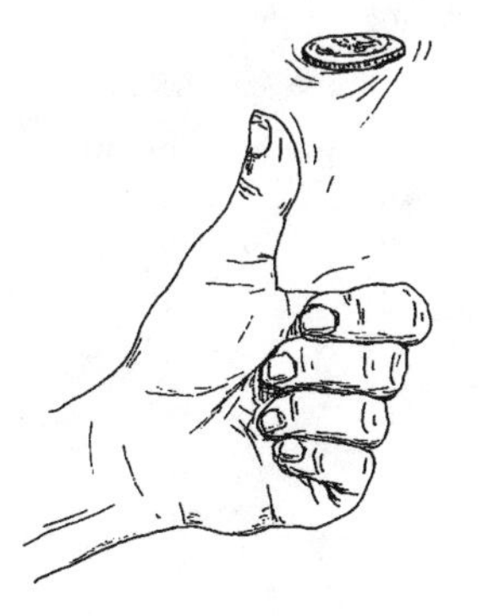

## 50 A New Variant of Heads or Tails

Two brothers are playing heads or tails, with only one coin. The older brother offers a change to the rules to his little brother: "It really is no fun if the first toss determines the winner. I propose that we toss the coin repeatedly until three successive throws form the sequence tail-heads-heads (THH), or the sequence heads-heads-tails (HHT). If the sequence THH appears first, I win; if not and it is the sequence HHT that appears first, you win." The younger brother refuses immediately, claiming the game isn't fair.

Why?

The older brother protests: "It is clear that the probability that THH appears in three tosses is equal to the probability that HHT appears in three tosses, which is $\frac{1}{8}$. Thus, the average time for them to appear are the same: the sequences appear on average after eight throws. The game is thus fair." What mistake(s) is the older brother making?

*For example, if the coin lands on THTHH, the game stops because the sequence THH appeared and the older brother wins the game.*

## 51 The Little Brother's Revenge

This riddle is a generalization of the previous one. The younger brother tells his older brother that he is willing to play under certain conditions: his older brother starts by choosing his sequence of three outcomes, then he chooses a different sequence of outcomes for himself. The first sequence to appear determines the winner.

Show that, whatever sequence the older brother chooses, the younger brother can always choose a sequence which makes him win on average.

# 52 Unbreakable Plates

A plate maker wants to test the solidity of his products. To do so, he takes two of his plates and goes to the foot of a 100-story building. He is trying to determine the limit floor, the lowest floor from which one of his plates breaks if dropped. He wants to make this determination in the fewest number of drops.

With the best possible strategy, in how many drops can he determine the limit floor with certainty?

*Both plates have the same solidity and do not wear during the drops. Following this method, in the worst case,* 100 *drops are made and only one plate is used. However, if two plates are used, we need to drop only from the even-numbered floors (*2, 4, 6, *etc.) until the plate breaks. Once the plate breaks, we then have to go to the floor immediately below it to find the limit. This strategy requires only* 51 *drops in the worst case, but it is still not optimal.*

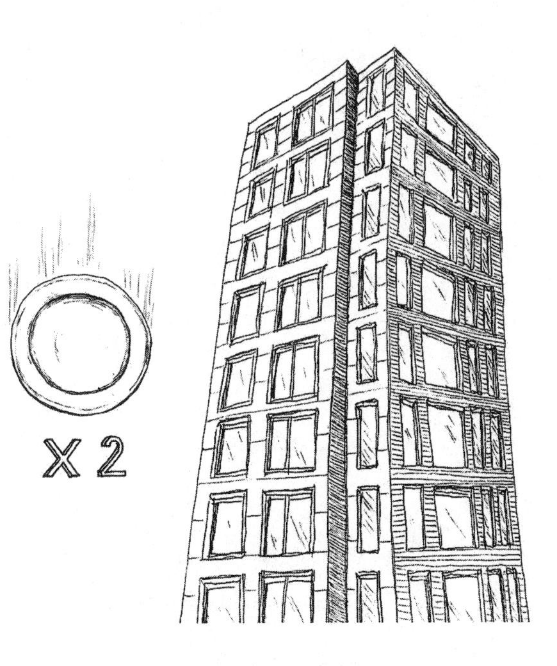

# 53 The Cable Around the Earth

We model the Earth as a sphere with radius $R = 6{,}400\,\mathrm{km}$. A taut cable is tightened at the equator. We lengthen the cable by 1 meter. We pull the cable outward at one point: it remains in the equatorial plane but is taut so that it is as high as possible at this point.

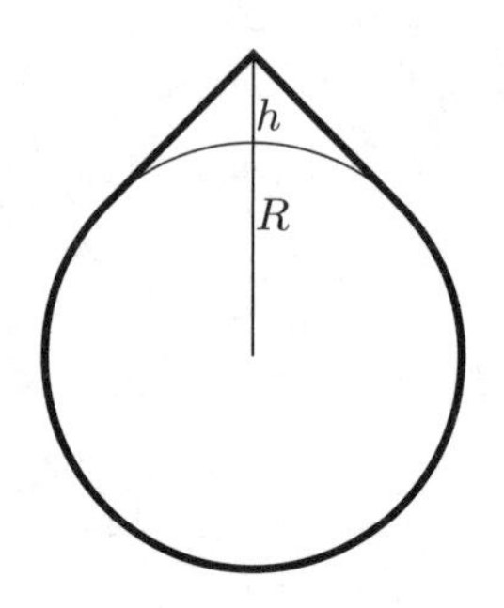


Determine the height ($h$) of the cable at this point.

# 54 The Spaghetti

A bored young man has before him a plate containing 100 spaghetti. He decides to knot them to form loops. To do this, he starts by taking two spaghetti ends that protrude, at random, and knots them together. He repeats this operation until all the spaghetti have been knotted.

At the end of this process, how many loops has he made on average?

# 55 Swimming Pool Lockers

At the swimming pool, a little girl is playing with the lockers. There are 100 lockers, numbered from 1 to 100, all of which are closed at the beginning. She makes a first pass by opening all lockers whose number is a multiple of 2. When she arrives at the $100^{\text{th}}$ locker, she comes back to locker 1 and makes a second pass. This time, she chooses all the lockers whose number is a multiple of 3; if the chosen locker is closed, she opens it, and if the locker is open, she closes it. She makes another pass in this manner for multiples of 4, 5, 6, and so on until her last pass, which is for the multiples of 100.

After all these passes, which lockers are closed?

*During her first pass, the girl opens the lockers* 2, 4, 6, *and so on through* 100 *and leaves the others shut. During the second pass, she opens locker* 3, *closes locker* 6, *opens locker* 9, *closes locker* 12, *and so on.*

# 56 Met at a Party

Show that in any group of six people, there are at least three who know each other or three who don't know each other.

*Of course, the relation of knowing someone is symmetric: if A knows B then B knows A. If one person knows everybody but the others know only this person, it suffices to consider a group of three people excluding the one who knows everyone to be a group of three people who do not know each other. However, in this case, it is impossible to find a group of three people who know each other.*

## 57 The Rats and the Bottles

A king owns precisely 1,000 bottles of wine. Having dismantled a plot against him, he knows that exactly one bottle is poisoned. His butler, charged with determining which one it is, has at his disposal some rats which he can have taste the different bottles. If a rat drinks of the poisoned wine, it dies the next day. The king is impatient and wants to be certain of which bottle is poisoned as early as tomorrow.

What is the minimum number of rats the butler needs?

*If each rat tastes one bottle, there has to be 999 rats to determine which bottle is poisoned. But the rats can also taste mixes from different bottles. For example, if the first rat tastes a mix of the first 100 bottles, he dies the next day if and only if one of these 100 bottles was poisoned. Note, you will have to prove that the number of rats proposed is in fact minimal.*

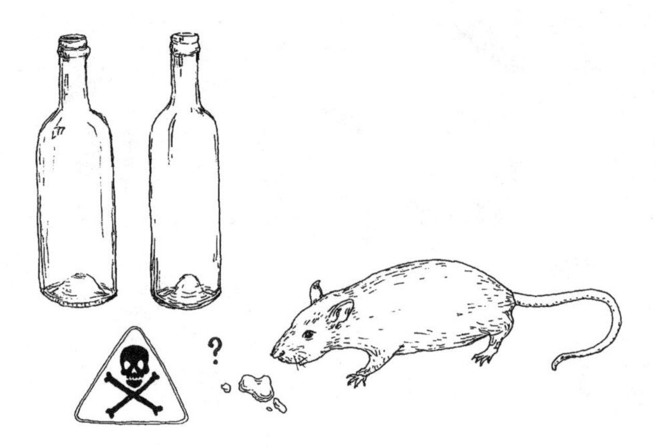


## 58 The Collection of Cards

A little girl wants to collect a new deck of cards that has just been released. There are 100 different cards to collect. The girl buys one card each day, but she cannot choose which card she buys: the card that she buys is chosen at random among those 100 cards.

How many days will it take the girl on average to complete her collection?

# 59 The Saint Petersburg Paradox

A banker offers the following game to his customer. He throws a fair coin. If it lands on heads, he gives his customer \$1 and the game stops; if not he tosses the coin again, and if it lands on heads gives \$2 to the customer and the game stops; if not he tosses the coin again, and if it lands on heads gives \$4 to the customer and the game stops, and so on. In other words, the banker tosses the coin until it lands on heads and gives $\$2^{n-1}$ to his client if it landed on heads after the $n$th throw.

What is the price of this game?

*In this game, the customer always wins. Finding the price of the game means finding the fee that the customer has to pay to his banker before he begins to toss the coin so that the game is fair, which is to say that neither of the two players has an advantage in the long run.*

# 60 The Chocolate Bar

Two mathematicians are sharing a bar of chocolate whose lower left square is poisoned. They impose the following rule: when someone wants to take a square of chocolate, they must take the whole upper right corner with it. For example, on the figure below, the first took square 1, then the second took square 2.

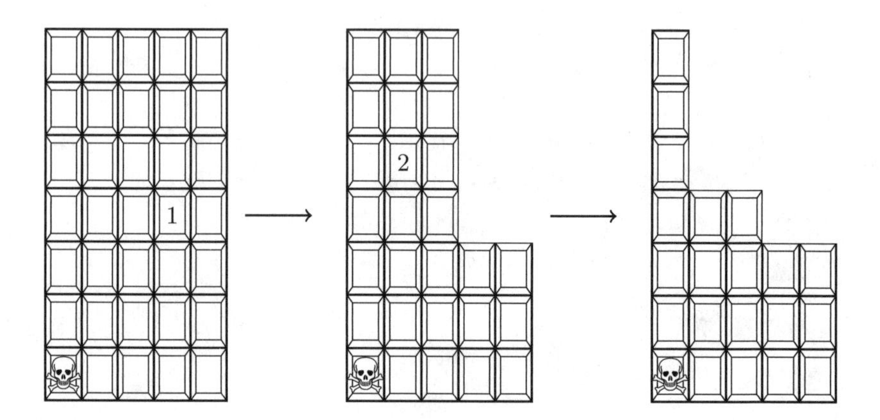


Each takes their turn to take a square and the upper right corner that comes with it. The one who is forced to take the poisoned square (when it is the only square left) is declared the loser!

Show that there is a winning strategy for the player who begins.

*Of course, we don't ask for the explicit winning strategy: this is an open problem!*

## 61 Tic-Tac-Toe

Two students invent a small game. They take turns choosing a number between 1 and 9 that hasn't been chosen before. The first who has three numbers whose sum is 15 wins.

Is there a winning strategy, and, if yes, for whom?

*For example if they choose in turn* 2, 1, 5, 8, 6, 3, 9, 4, *the second student wins since among the four numbers he chose* 1, 8, 3, 4, *the last three have a sum equal to* 15. *Note that there may be a draw, if all the numbers are chosen but no player has three that sum to* 15.

## 62 Ant vs. Car

A race car is 1 meter away from a pole, to which it is connected by a taut elastic. On this elastic, by the pole, is an ant. The car starts suddenly: it instantly accelerates to 200 km/h. At this precise moment, the ant starts moving on the elastic toward the car, which is moving away from it. On the ground the ant would move at 1 m/min, so it has a speed relative to the elastic of 1 m/min.

Will the ant catch up with the car? If so, after how much time?

*The car moves indefinitely at* 200 km/h *and the elastic is assumed unbreakable: it can extend homogeneously indefinitely.*

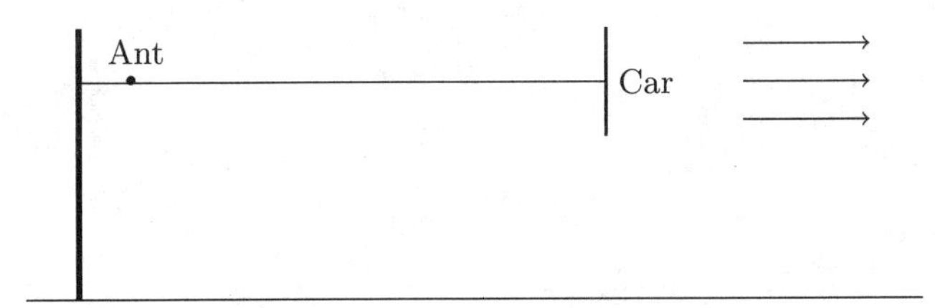

# 63 The Squirrel and the Nuts

A squirrel collects 11 nuts. Back in its hideout, it weighs up its spoils in many ways. It notices that whichever nut it sets aside, it can always separate the 10 others into two groups of five nuts with the same weight. We assume that all the nuts have a mass that, when expressed in grams, is an integer.

Show that all the nuts have the same mass.

*Assume that the nuts all have the same mass. In this case, it is evident that, whatever the nut the squirrel sets aside, it can always separate the* 10 *remaining into two groups of five nuts with the same mass. Assume now that the nuts respectively weigh* 1 *gram,* 2 *grams, ...,* 11 *grams. If the squirrel sets aside the nut that weighs* 2 *grams, it can separate the* 10 *remaining nuts into two groups of five with the same mass: on one side the nuts weighing* 1 *gram,* 4 *grams,* 6 *grams,* 10 *grams, and* 11 *grams, and on the other those weighing* 3 *grams,* 5 *grams,* 7 *grams,* 8 *grams, and* 9 *grams. However, if it sets aside the nut weighing* 1 *gram, we can show that it is impossible to separate the* 10 *remaining nuts into two groups of five with the same mass. Thus, the distribution of masses of the nuts proposed in this particular case does not work. One must show that the only distributions that work are those in which the nuts have the same weight.*

# Level 3

## 64 The Triangular Cake

A husband, to celebrate the birthday of his beloved, orders a triangular cake from the best baker in town. In the meantime, he crafts a triangular box in which to beautifully present the cake. When he goes to pick the cake up from the baker, the husband is disheartened! He realizes that the baker adhered to the measurements that were given but made the exact symmetric version (see figure below). Therefore, the cake doesn't fit in the box. The husband thinks for a moment, then realizes he needs only to make two cuts in order for the cake to be cut into pieces which, once rearranged, will perfectly fit the box.

How does he do it?

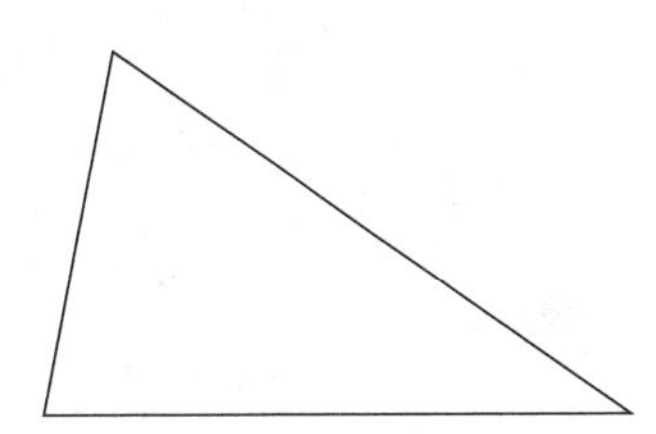

The cake (from above)

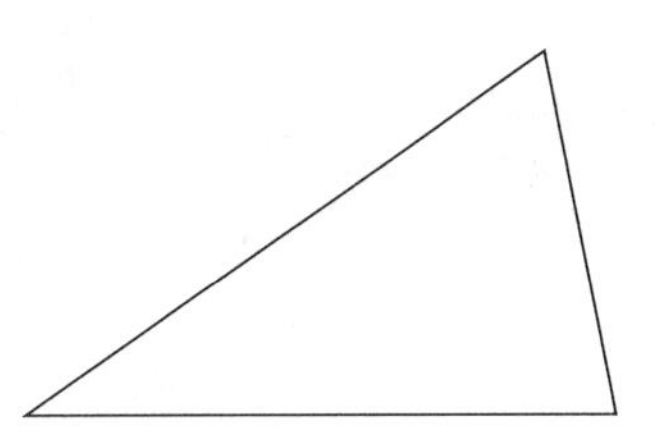

The box (from above)

*Of course, the precise triangular shape of the cake is a priori arbitrary. One can't stack parts or flip them over.*

## 65 A Fair Share

To share a pizza equitably between two people, the first person cuts the pizza, and the second chooses their slice.

Determine, under the same principle, a fair way to share a pizza between $n$ people.

## 66 The Lonely Soldier

Eleven soldiers are posted in a field. The orders are clear: each soldier must watch the soldier closest to him.

Show that there is at least one soldier who is watched by none.

*To remove ambiguities, we assume the distances between pairs of soldiers are all different.*

# 67 The Booty of the Thirteen Pirates

Thirteen pirates decide to gather their shared booty into a chest. They want to design a system that allows them to open the chest as soon as a majority of pirates want to open it, and not otherwise. To do so, they call a blacksmith, who places a certain number of padlocks on the chest. The chest cannot be opened unless every padlock is opened. The blacksmith makes multiple sets of keys for each padlock. He distributes them among the pirates in such a way that each pirate has some number of keys, without having all of them.

To design such a system, what is the minimum number of padlocks that the blacksmith must use?

*For example, if there are only three pirates, three locks suffice: it is sufficient to give the first pirate the keys to padlocks* 1 *and* 2*, to give the second pirate the keys to padlocks* 1 *and* 3*, and to give the third pirate the keys to padlocks* 2 *and* 3*. In this way, none can open the chest on their own, but when two pirates come together, they can open it.*

## 68 Reference Weights

A set of Roberval scales is sold with a set of 10 reference weights.

What masses must the 10 weights have in order to weigh as many consecutive (integer) mass objects as possible?

*For example, with only three weights, one weighing* 1 kg, *one weighing* 2 kg *and one* 5 kg, *it is possible to weigh any object between* 1 kg *and* 8 kg. *Do note, we ask not only for an optimal set of weights, but also the proof of its optimality.*

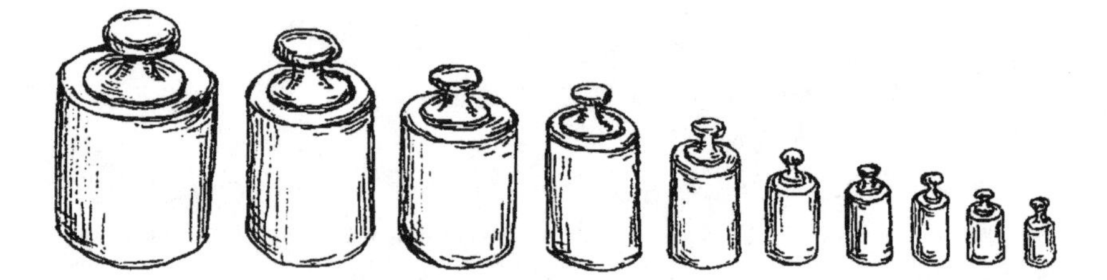

## 69 The Elephant and the Bananas

A banana merchant has a stock of 3,000 bananas that he would like to sell at the market of an oasis situated 1,000 kilometers away. To carry the merchandise he has an elephant that consumes one banana per kilometer and cannot carry more than 1,000 bananas at once.

At most, how many bananas can he transport to the oasis?

*First off, we notice that if the merchant does the trip without stopping, loading up his elephant as much as possible, he will arrive at the oasis with no bananas. It is thus necessary to make stops: the idea is to drop off some merchandise on the way, come back to pick up more bananas, and then pick up the bananas left on the way. Note, we ask not only for an optimal strategy, but also the proof of its optimality.*

## 70 The Very Hungry Termite

A termite is eating a wooden cube made up of 27 smaller cubes, as shown on the following diagram. It begins with a small cube of its choice, then moves to an adjacent cube (it cannot move diagonally) until it has eaten every one of them.

Can the termite finish its meal by eating the central cube?

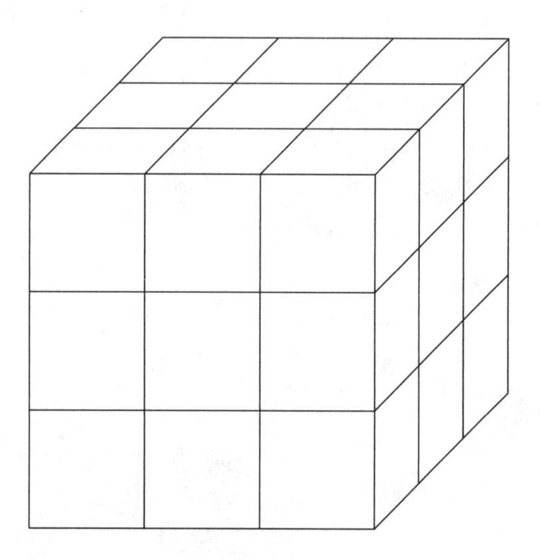

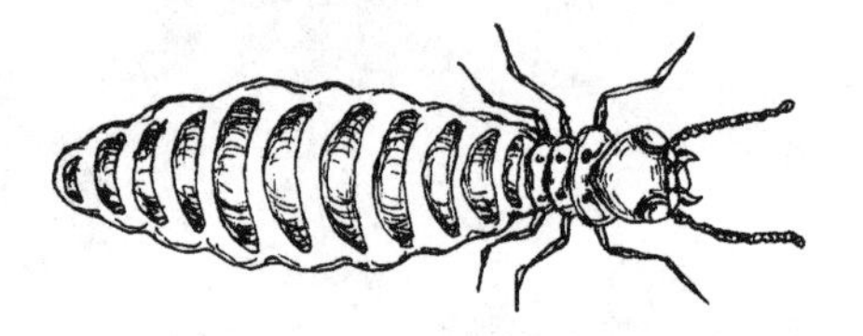

# 71 The Principal and the Lightbulb

A principal offers a challenge to his 100 students. "I will call you one by one into my office, perhaps several times, until one of you can assure me that everyone has been through my office, in which case you all win. So that this is possible, you will be able to, when called, either turn on or off the light in the hallway leading to my office. You will not be able to communicate, so you will not know who came in before you, or even how many students came in before you. I will keep calling you in as long as none of you has assured me that all of you have come in. If one of you tells me everyone has been in my office and that is not the case, you all lose."

How can the students win this challenge?

*It is not specified whether the lamp is on or off at the beginning. Do note that students can enter the office several times before one of them announces that they have all been in. If, for example, each student has been in the principal's office* 50 *times before one of them declares that they all have been in, they win nonetheless.*

# 72 The Principal and the Drawers

The principal offers another challenge to his students. The students are numbered from 1 to 100, and they each know their own number. Then, they are called one by one and only once into the office of the principal, where there is a chest with 100 drawers numbered from 1 to 100. The principal has randomly placed into each drawer a piece of paper with the number of a student. Each drawer contains exactly one number, and each number is contained by exactly one drawer. Each student must find their own number. Each student can open a maximum of 50 drawers. The students cannot communicate during the challenge, do not know who was called before them, and cannot move the papers in the drawers. All the students win if every one of them finds their number. They can agree on a strategy before the challenge starts.

Find a strategy that gives students at least 30% chance of winning.

*First of all, regardless of the chosen strategy, it is clear that the first student will only have a 50% chance of finding his number. The probability that students win is thus less than 50%. At first glance, each student only has a 50/50 chance of finding their number and the probability of winning is only $2^{-100}$, which is to say near 0!*

## 73 The Principal and the Seven Hats

The principal, who isn't lacking in imagination, invents a new test for his poor students. He calls in seven students and tells them about the fiendish new challenge. "You will be placed in a room, and I will place a hat on each of your heads. There are seven possible colors: yellow, orange, red, blue, green, brown, and purple, but I reserve the right to not use all these colors. For example, I may place three yellow hats and four purple hats, should I so desire. Then, I will leave you a few minutes to look at the colors of the hats of your friends. Afterward, I will ask you each in turn to whisper in my ear, so that the others don't hear, the color you think your hat is. If at least one of you guesses the color of your own hat, you all win; otherwise all of you lose."

What strategy can the students use so that they win this challenge?

*Of course, no student sees the color of their own hat, and the students cannot communicate during the test.*

## 74 The Principal and the T-Shirts

With the same seven students, the principal has found a variant of the previous game. "This time there will be a different real number written on each hat. As before, you can see the hats of others but not your own. Each of you must then choose a T-shirt that is either black or white and wear it. Afterward, you will be placed in increasing order of the numbers on your hats. You win if the T-shirts are of alternating colors."

What strategy can the students use to win this challenge?

*Students choose their T-shirts all at once and in such a way that no student can draw any information from the color of the T-shirt of another student.*

# 75 Die Hard

A wine maker wants to draw 1 liter of wine from his barrel, which contains as much wine as necessary. He has on hand only a pitcher of $p$ liters and a pitcher of $q$ liters. They have no markings. We can suppose that $p$ and $q$ are coprime.

How can he obtain a liter from these two pitchers?

*One could start by solving the simple case appearing in a famous blockbuster:* $p = 3$ *liters and* $q = 5$ *liters.*

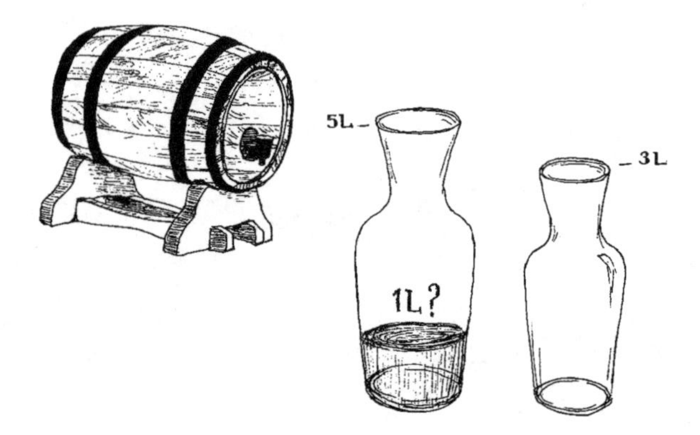


# 76 The Rats and the Bottles (2)

Another king owns precisely $p$ bottles of wine. Having also dismantled a plot against him, he knows that exactly one bottle is poisoned. His butler, charged with determining which one it is, has some rats which he can have taste the different bottles. If a rat drinks the poisoned wine, he dies the next day. This king is patient and gives his butler $q$ days to find out for certain which bottle is poisoned.

What is the minimum number of rats that the butler needs?

*Like in riddle* 57*, the butler will have each rat taste a mix of bottles. The difference is that the next day, he will be able to reuse the rats that survived to have them taste different mixes; this protocol can be repeated for* $q$ *days. Beware, one will have to prove the number of rats is in fact minimal.*

## 77 Where Is the Robot?

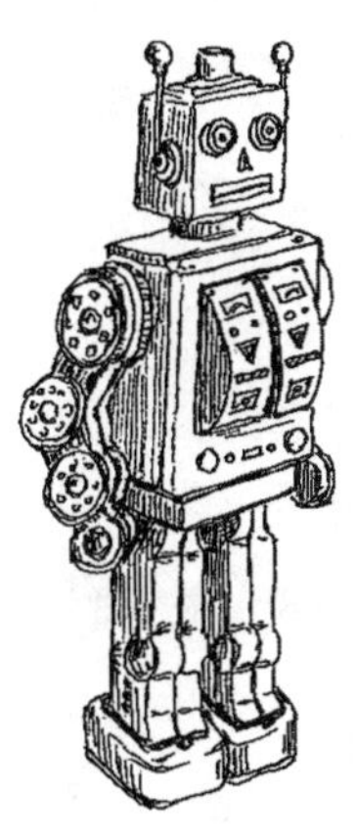

A robot is moving along the axis of integers. At $t = 0\,\text{s}$, he is at integer $m$. Each second, he moves with a constant step size $p$, which is a whole number. For example, if $m = 5$ and $p = -2$, at $t = 1\,\text{s}$ the robot is at 3, at $t = 2\,\text{s}$ the robot is at 1, and so on. We know neither $m$ nor $p$, but at each second, we can ask, "Is the robot on $n$?" where $n$ is an integer and the answer is a binary response ("yes" or "no").

Design a strategy that can be used to find the robot's exact location after a finite amount of time.

## 78 The Seven Bridges of Königsberg

The town of Königsberg (today Kaliningrad, Russia), through which flows the river Pregolia, had seven bridges arranged in the following pattern:

If someone carefully chooses the starting point of their walk, is it possible for them to walk around town by crossing each of the seven bridges exactly once?

# 79 Linking the Edges

Consider the following figure:

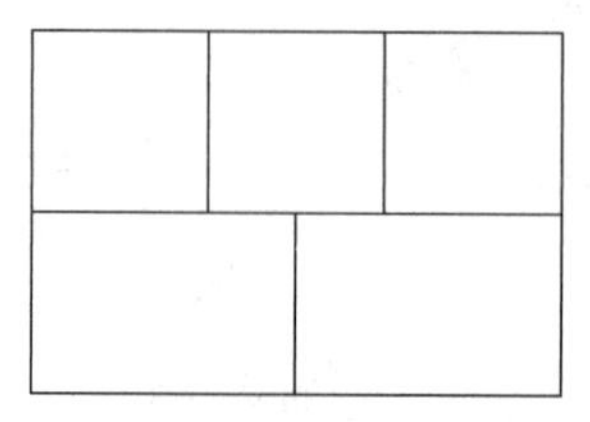

Is it possible to draw a curve passing exactly once through each edge without lifting the pen off the page?

*For example, in the attempt below, the bold edge has not been crossed by the curve, so this one is a failure.*

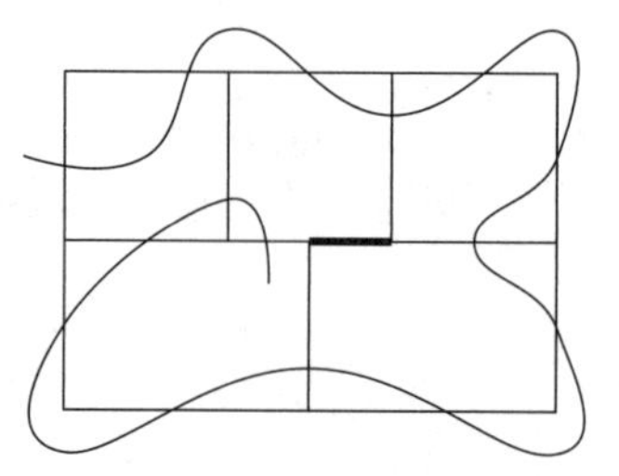

# 80 A Magic Trick

Alice and Bob perform a magic trick with a special pack of cards containing 124 cards numbered from 0 to 123. A spectator chooses five cards. He gives them to Bob, who examines them, removes one of his choosing, then places the remaining four cards on the table after having ordered them as he pleases. Then Alice, who was standing aside, looks at the remaining four cards, and determines which card had been removed!

How do they do it?

*The only way for Alice and Bob to communicate is through the order of the remaining cards.*

## 81 How to Tile One's Bathroom (2)

The floor of a bathroom is a square consisting of 49 tiles. Following some work, the upper left corner tile is removed. The bathroom floor thus takes the shape below:

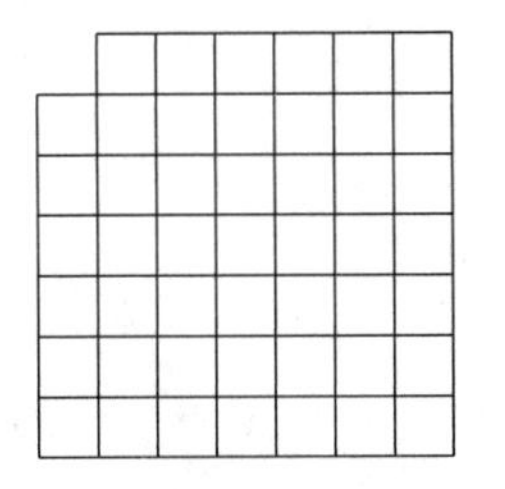

We wish to cover it with tiles in the shape of an L, which we can also invert:

Is it possible?

## 82 How to Tile One's Bathroom (3)

The floor of a different bathroom is a square consisting of 64 tiles. Following some work, the upper left corner tile is removed. The bathroom floor thus takes the shape below:

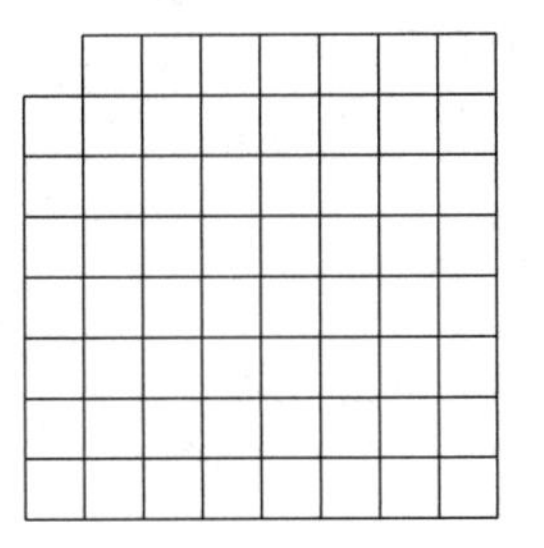

We wish to cover it with tiles in the shape of an elongated domino:

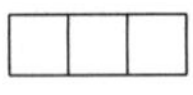

Is it possible?

# 83 The Mad Postman

In a small village, a road containing 100 houses loops back on itself: thus, house number 100 is next to house number 1. A mad postman delivers the mail. He begins by house number 2 and moves forward distributing the mail to every second house, leaving an empty mailbox each time. Thus he begins by distributing the number $2, 4, \ldots, 100$, then to house number 3, house number 7, house number 11, and so on.

What house receives its mail last?

*For example, with six houses, the mailman distributes the mail to house numbers* 2*,* 4*, and* 6*, then to number* 3 *(he skips number* 1 *and ignores number* 2*), number* 1 *(he ignores number* 4 *and skips number* 5*), and finally finishes with number* 5*.*

# 84 Mind the Dog!

A tourist who ignored the titular sign is swimming in a circular lake. On the edge of the lake is the dog that is running after her and is compelled to move in a circle along the shore: at any point in time, the dog tries to be as close as possible to the unlucky tourist.

To escape the dog and get out of the lake, what is the minimum speed at which the tourist should swim, and in which direction should she do so?

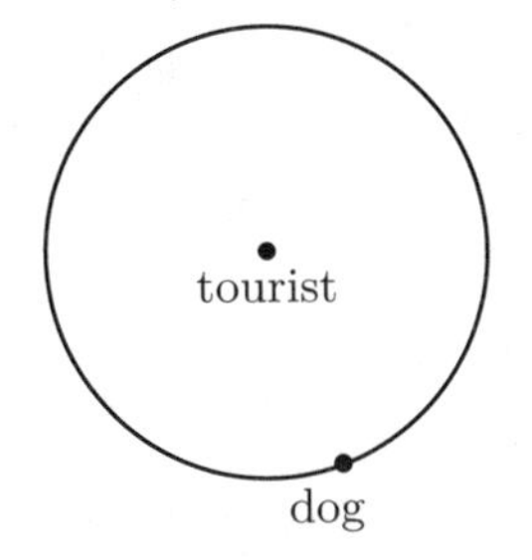


*The tourist can outrun the dog on land, so if she reaches the edge of the lake before the dog, she can escape. For example, if the tourist swims at a speed greater than* $\frac{v}{\pi}$*, with* $v$ *the speed of the dog, she can escape. Indeed, it suffices for her to swim to the center of the lake, then swim to the point across the lake from the dog. Not only can she escape at a speed strictly below* $\frac{v}{\pi}$*, she can even escape at a speed strictly less than* $\frac{v}{1+\pi}$*. Note, you will have to prove that the speed limit proposed is indeed minimal.*

# 85 The Two Envelopes

A banker, who is in a very good mood, offers the following game to his customer. He fills two checks in customer's name with two different sums. He places each check in an envelope. The two envelopes are identical. If the customer has to choose one of the envelopes and open it, then he gains on average $\frac{1}{2}(M_1 + M_2)$ where $M_1$ and $M_2$ are the sums on the checks. The banker offers him something different: the customer must choose one of the envelopes, open it, look at the sum on the check it contains, and choose whether to keep the sum or open the other envelope to cash the other sum.

Is there a strategy that allows the customer to gain on average more than $\frac{1}{2}(M_1 + M_2)$ ?

*It seems like there wouldn't be, since the customer has no information about $M_1$ and $M_2$ other than that both values are positive and different from one another. Nonetheless...*

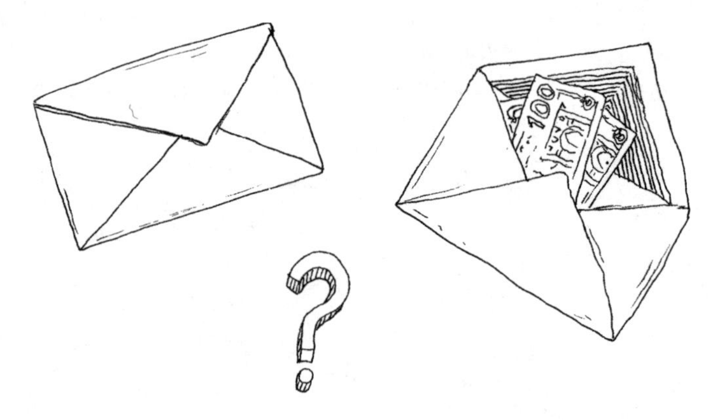

# 86 The Two Envelopes (2)

This time, the banker tells his customer he has put a sum at random in one envelope and twice that amount in the other; both envelopes are identical. The banker hands the customer the envelopes, and the customer must choose one. He arbitrarily picks the one on the left. That is when the banker claims to prove that the customer would have been better off choosing the envelope on the right! Here is his reasoning:

"Let $x$ be the amount in the left envelope (we don't know $x$ since it hasn't been opened). The sum in the envelope on the right has a 50/50 chance of being $2x$, and a 50/50 chance of being $x/2$; thus, the envelope on the right contains on average

$$\frac{1}{2} \times 2x + \frac{1}{2} \times \frac{x}{2} = \frac{5x}{4}$$

which is strictly more than $x$; it is thus beneficial to change envelopes!" This is absurd; since both envelopes are identical, there is no better way to choose one of the two arbitrarily. Hesitating in turn between the left and the right envelopes will not increase the average gain.

Where is the mistake in the banker's reasoning?

# 87 The Squirrel and the Nuts (2)

A squirrel collects 11 nuts. Back in its hideout, it weighs up its spoils in many ways. It notices that whichever nut it sets aside, it can always separate the 10 others into two groups of five nuts with the same weight.

Show that all the nuts have the same mass.

*We no longer assume that the weights are integers like we did in riddle 63.*

## 88 The Monkey with a Typewriter

A monkey types on random keys of a typewriter that has only 26 keys, one per letter of the alphabet. It types at random, with a constant speed of one letter per second. It favors no letters: all letters at any second have a $\frac{1}{26}$ probability of being typed. Let $t_1$ be the average time it will take the monkey to type "abracadabra" and $t_2$ be the average time it will take the monkey to type "abracadabrx".

Compare $t_1$ and $t_2$.

*The word "abracadabra" has 11 letters, so it has a probability of $26^{-11}$ of being typed in 11 seconds. It is the same for the word "abracadabrx" which could lead one to think that their mean occurrence time is the same. It is not true; see riddle 50 for more details on this type of erroneous reasoning.*

# 89 A Rectangle with Integer Sides

A big rectangle is partitioned into smaller rectangles each having at least one side whose linear measurement is an integer, like in the example below:

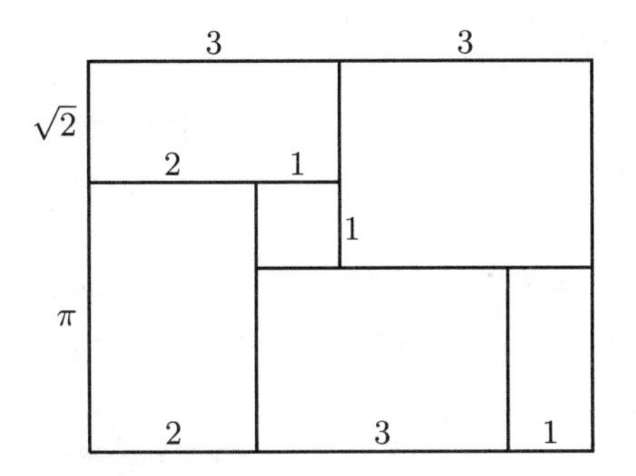


Show that the length or the width of the big rectangle is an integer. If you can, show that both are integers.

*For example, on the figure, the length of the big rectangle is an integer (but not its width).*

# 90 Finding the Right Wire

An electrician has connected two houses with 10 wires. The wires start from the first house, where their starting points are numbered from 1 to 10, and arrive at the second house, where their end points are labeled A to J. The electrician got confused while numbering the wires and can't remember which starting point corresponds to which end point. To figure things out, he has a lamp, a battery, and as many screw terminals as he wishes in order to connect the wires.

How can he, by starting from the first house and with only one round trip, find which pairs of start and end points match?

*He can, for example, link with screw terminals start points* 1 *and* 2*, then go to the second house and plug the battery into end point $A$ and the lamp into end point $B$. If the lamp turns on, then starting point* 1 *corresponds to end point $A$ (which means they are on the same wire) and point* 2 *corresponds to point $B$, or vice versa that point* 1 *corresponds to point $B$ and point* 2 *corresponds to point $A$. If the lamp does not turn on, he knows that the pair of end points $A/B$ does not match the pair of start points* 1/2.

# 91 Making Friends

It is back-to-school day. Among the 30 students of a classroom, some know each other from the previous year. To help the students meet more of their fellow classmates, the teacher wishes to divide the classroom into two groups, not necessarily of the same number of students, so that each student knows fewer people inside their group than in the other.

How can this be done?

*Of course, the relation of knowing someone is symmetric: if student $A$ knows student $B$, then student $B$ knows student $A$. On the other hand, it need not be transitive: it is possible that student $A$ knows student $B$ and that student $B$ knows student $C$ without student $A$ knowing student $C$. The teacher knows all these relations perfectly and must divide the class as a result.*

## 92 Red or Black?

Alice offers Bob a small game with a standard 52-card deck. Alice will reveal to Bob the 52 cards one by one. When he chooses, Bob will have to predict that the next card is red. Bob will be able to consider the colors of the revealed cards to choose the right moment to make his prediction, but he must make the prediction at some point, even if all the red cards are gone.

Is there a strategy that allows him to have more than a 50/50 chance of success?

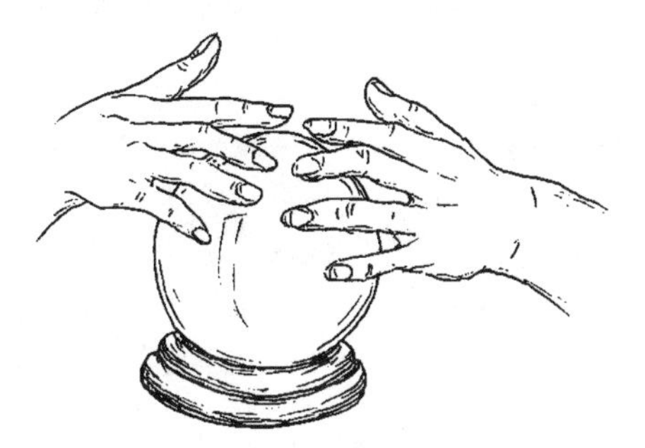

## 93 Finding the Cable

A farmer has a square field whose side is 1 km long. He knows a cable passes under his field and that the cable is straight. For example, the cable can be located as shown in this picture:

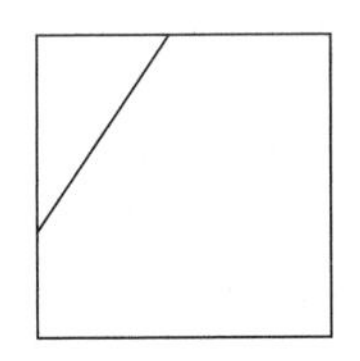

To find the cable, he considers digging trenches. He can, for example, dig both diagonals of the square. Using this method, he is sure to intercept the cable. However, this requires digging a length of $2\sqrt{2} \simeq 2.83$ km.

Find a strategy in which the farmer will never have to dig more than 2.64 km.

## 94 Lukewarm Earnings

A large industrial corporation is holding a press conference to announce its results. The CEO proudly proclaims: "Note that during every consecutive 8-month period, the company has turned a profit." A stockholder interrupts him: "Yes but during any consecutive 5-month period, the company has posted a loss."

How many months are covered by the earnings?

A keen reader can also take up the general case by replacing 5 by $p$ and 8 by $q$, with $p$ and $q$ coprime integers.

*We will assume that profits and losses are strict and that* 0 *is neither a profit nor a loss. For example, these earnings could cover the last* 8 *months. Indeed, it suffices that the company earned* $+1$ *at each month except the fourth one, when it earned* $-6$*. The company is indeed in the green during any consecutive* 8*-month period (there is only one, which earned* $+1$*), and it is in the red on any consecutive* 5*-month period (*$-2$*).*

## 95 Not Your Usual Game of Chess

Alexei offers a challenge to Ivan and Dimitri: he places 64 coins on a chessboard, one on each square, either on heads or tails. He shows the board to Ivan and selects a square that will be the secret square. Ivan must flip exactly one coin, although he can choose which coin to flip. Once the coin has been flipped, Ivan leaves and Alexei presents the chessboard to Dimitri, who is seeing the board for the first time. Dimitri must then determine, simply by looking at the chessboard, which square was the secret square.

What strategy can Ivan and Dimitri put into place to beat the challenge?

Extra credit: show that putting a winning strategy in place is possible only if the board has a number of squares equal to a power of 2.

## 96 The Principal and the Real Hats

Consider a countably infinite number of students placed on a staircase with infinitely many steps: step number 0 is at the very top, followed by step number 1, step number 2, and so forth. The staircase descends infinitely. For each student, the principal chooses a hat that is inscribed with a real number. We know nothing of the sequence of real numbers formed in this way; the principal chooses the numbers as he pleases. Each student knows on which step they are standing but can only see the hats of students on lower steps. Thus, the student at the top of the staircase sees the hats of all other students, and each student sees an infinite number of hats. The students must announce the real number on their heads all at the same time. They may agree on a strategy before the challenge starts.

Find a strategy ensuring that only a finite number of students will ever get their own number wrong.

*Unlike in riddle 39, students cannot get any information by listening to what others say since they must all announce their number at the same time. If, for each student, the principal chooses a number independently and uniformly at random between 0 and 1, it would seem that no student would succeed at guessing their number; yet, almost all of the students succeed.*

## 97 The Principal and the Hats (2)

This riddle is a generalization of riddle 39. This time there is a countably infinite number of students. They are placed on a staircase with infinitely many steps: step number 0 is at the very top, followed by step number 1, step number 2, and so forth. The staircase descends infinitely. The principal places either a black hat or a white hat on each student's head. We know nothing of the distribution of colors; for example, it is possible for the principal to hand out only white hats. Each student can see the hats only of students below them. Thus, the student at the top of the stairs can see the hats of all other students and each student sees infinitely many hats. The principal asks the student at the top about the color of her hat, to which she must respond by saying "white" or "black," and all the other students hear her response. Then, it is the turn of the student on step 1 to be asked, then of the student on step 2, and so on until all students have been asked. (One could imagine that the response time decreases exponentially.) Students may agree on a strategy before the start of the challenge.

Find a strategy that ensures that at most one student guesses wrong.

*It's clear that the student at the top has no information on the color of her hat, so she cannot be sure to find the color of her hat. Like in riddle* 39, *her mistake must be useful to the others in guessing since they must make no mistakes.*

# 98 One Last Riddle with the Principal

One hundred students are called one by one, only once each, into the office of the principal where there is a desk of drawers with an infinite number of numbered drawers: there is drawer number 0, drawer number 1, drawer number 2, and so on. In each drawer, the principal has left a piece of paper with a real number of his choosing. We know nothing of the sequence of real numbers thus created; it was chosen arbitrarily by the principal. Although no student is allowed to open every drawer, there is no other restriction on the number of drawers a student opens or on the order in which they open the drawers. Eventually, the student must stop and predict the real number of a drawer that they did not open. After the student makes the prediction, all the drawers are closed again, and the next student is called in. Students may not communicate during the challenge, but they can agree on a strategy beforehand.

Find a strategy which allows at least 99 students to guess correctly.

# 99 The Lax Captain

During the review of troops, soldiers are generally all lined up. A captain permits his solders to arrange themselves as they wish under the condition that, to not infringe on military decorum too much, each line of soldiers contains at least three soldiers.

Show that this will not change the usual lineup: the soldiers will all necessarily form a single line.

*Formulated differently, this problem consists in showing that if a finite set of points in a two-dimensional plane are such that any line passing through two points passes through another point of the set, then all the points are lined up.*

# 100 Peg Solitaire

One has an infinite grid on which marbles are placed in the lower half-plane delimited by a line.

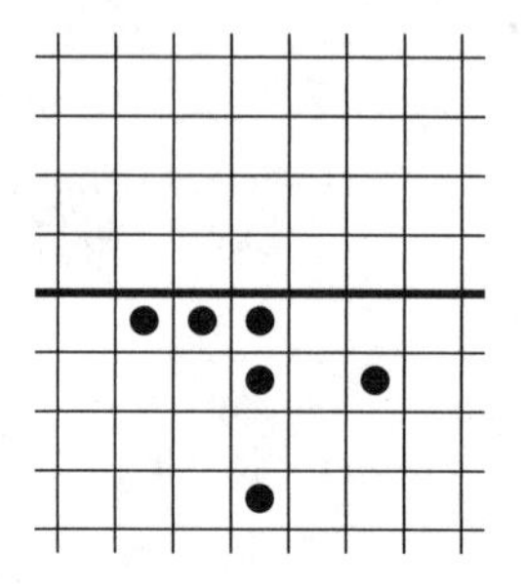

Marbles may move like in the game of peg solitaire, which means that a marble can "jump" (horizontally or vertically but not diagonally) over another peg, which is then removed, like so:

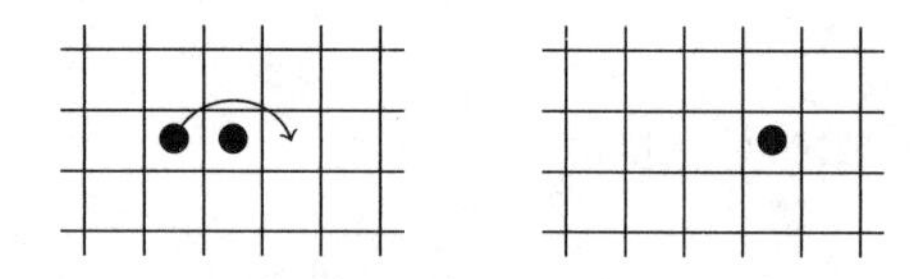

The goal is to send a marble as high as possible beyond the line.

What is the maximum height one can reach?

*For example, on the above grid, it will not be possible to go more than two squares above the line. However, with a good configuration of the marbles below the line, one can do better.*

# Part II

# Hints

**1** It is clear that opening four links is sufficient, but is it optimal?

**2** The easiest approach is to begin by thinking about what happens when there are only two ants dropped on the stick, then only three.

**3** Determine the price of a cake by thinking about the total cost of the cakes.

**4** The easiest approach is to reason about the mass of the fry, which remains constant when the diner removes ketchup.

**5** Possible starting points are south rather than north! Since the explorer is going 1,000 km south, he will end up on a parallel. What conditions on this parallel ensure that the explorer will return to his starting point when he leaves north?

**6** What happens when both ends of a vine are lit at once?

**7** One must add an A pill to the other three then cut the four pills in a clever manner.

**8** One can begin by assuming that the distance between his house and the university is 40 km and study the total round-trip time if his average speed on the round-trip is 40 km/h.

**9** What happens if there is a lot of current?

**10** The easiest approach is to consider the point of view of the midwife.

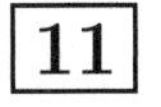

There is no computation to do.

**12** What happens during the last instance of the procedure that makes both mixes balanced?

**13** One must ask one of the guards a question that pertains to the other.

**14** The kangaroo has two possibilities to begin climbing the stairs: either he climbs the first step or jumps. How many ways can he finish climbing the stairs in each case?

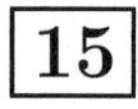 No need for computation.

**16** One might think there is missing information (distance from the school to the house, speed of the car, etc.) but all the necessary details are there. One must avoid jumping to computations: the solution is very short if approached correctly.

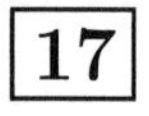 One could consider the small central cube.

 The sum is 89 and it is the opponent's turn to play. What happens?

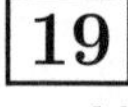 The idea is to work out how many living rooms the three painters would paint together in 30 hours.

 One could reason by contradiction.

**21** Plot the distance of the executive to her house as a function of time.

 The student places the first coin in the center of the board. What can he do afterward?

**23** The easiest approach is to work out the probability that no one shares a birthday with anyone else.

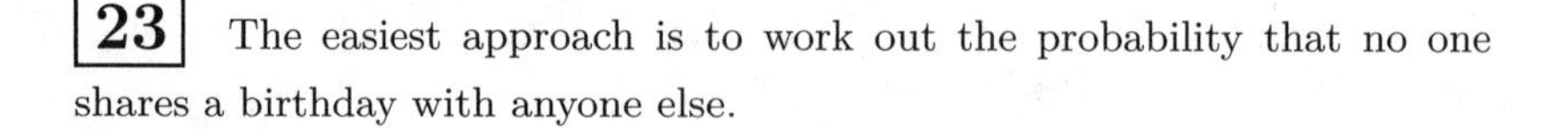

**24** What happens with one snake? With two snakes? With three? One will then be able to set up an induction (see reminder 5).

**25** One must use Bayes' rule (see reminder 8).

**26** The number of shots taken in each match must be different. We can thus give more or less weight to the percentages.

**27** One can consider the moment when the player's success rate goes over 75%.

**28** One must try all the possible combinations: what are the possible ages and what sums do they imply?

**29** One may proceed by trial and error. Another way, for the reader familiar with group theory, is to define four formal symbols:

- $a$ : wrapping the rope around the first nail clockwise
- $b$ : wrapping the rope around the second nail clockwise
- $a^{-1}$ : wrapping the rope around the first nail counterclockwise
- $b^{-1}$ : wrapping the rope around the second nail counterclockwise

Now, any way to hang the painting corresponds to a sequence of these symbols; the one given as an example in the statement is $ab$. One can try to find a sequence of symbols that translates to a solution. Two things must be identified: the effect of removing a nail on the sequence, and what corresponds to the painting falling.

**30** What happens if there are only two unfaithful spouses in the village? Only three? Only four? One must then reason by induction (see reminder 5).

**31** One must use Bayes' rule (see reminder 8) to compute the probability that at least one child is a boy given that at least one child is called Sophie. One should find $\frac{1}{2-\frac{\epsilon}{2}}$.

**32** Show that by changing doors, the candidate has a probability of winning equal to $\frac{2}{3}$.

**33** One can start by creating jigsaw puzzles with 6, 7, and 8 pieces, then deducing that for any $n \geq 6$ there is an $n$-piece jigsaw.

**34** If the Good kills the Ugly, then he ends up in a duel against the Bad, who has the advantage since he shoots first. On the other hand, if the Good shoots up into the air he will end up dueling the Bad or the Ugly, but being the first to shoot.

**35** One can reason by induction (see reminder 5).

**36** One mustn't reason card by card, but rather divide the number of cases in which the prediction is correct by the total number of cases. One must use binomial coefficients (see reminder 2).

**37** One can begin by placing weights 1 to 4 on one side and 5 to 8 on the other. The easiest way to find the solution is to use the method of comparing outcomes to hypotheses (see reminder 1).

**38** Some coins will have to be turned over.

**39** When there are only two colors, the top student must give global binary information. With seven colors it is recommended to use modular arithmetic (see reminder 6).

**40** One must separate the 10 coins into two groups of five coins.

**41** Consider the function that maps any point on the globe to the temperature difference between this point and the point on the opposite side of the Earth.

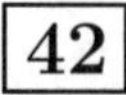

**42** For the second question, think of parity.

**43** No computation!

**44** What happens with two pirates? With three pirates? With four pirates? Keep thinking in this pattern.

**45** Check that if the baboon runs alternatively 0.5 km at a certain speed, then 0.5 km at another speed, he nonetheless has a constant average speed on any kilometer traveled.

**46** For each dwarf, one will have to put a different number of ingots on the scale.

**47** If we are given two points, in what zone must the third lie to create an acute triangle?

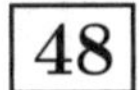

One could think of a chess board with its white and black squares.

**49** No computation!

**50** Separate the cases following the result of the first two throws. Can the younger brother win if the first two throws aren't HH?

**51** What makes the older brother win in the previous riddle is that the end of his sequence THH is the start of the younger brother's sequence HHT. One must try to use this advantage. By symmetry, there are only four cases left to examine, one of which has already been covered: TTT, TTH, THT, HTT.

**52** One could ponder the question backward by asking from how high we can drop the first plate if we only have at most $n$ trials.

We could use the following approximations for "small" $x$:

$$\tan(x) \simeq x + \frac{x^3}{3} \quad \cos(x) \simeq 1 - \frac{x^2}{2} \quad \frac{1}{1+x} \simeq 1 - x$$

**54** Let $b_n$ be the number of loops that the young man obtains on average on a plate of $n$ spaghetti and obtain a recurrence relation for $b_n$.

**55** Reason on the parity of the number of divisors of the number of each locker.

**56** Make a graph whose six vertices represent the six people and whose edges represent the relation of knowing each other or not. By considering one of these six people, begin by showing that they have three relations of the same nature.

**57** One must begin by asking oneself how many rats are necessary. For this, use the method of comparing outcomes to hypotheses, which is detailed in the reminder on the pigeonhole principle. Moreover, one must number the rats in base 2 (see reminder 1).

**58** Given that the girl already has $k$ distinct cards, denote $c_k$ as the average number of cards that the girl must buy and obtain a recurrence relation on $c_k$.

**59** Begin by assuming (realistically) that the banker doesn't have an infinite amount of money to give the customer.

**60** What happens if the mathematician going second plays a winning strategy?

**61** The title of the riddle is a hint!

**62** Add the absolute speed of the ant, 1 m/min, to the speed of the point of the elastic on which it stands.

**63** The easiest approach is to reason on the parity of the weights.

**64** One could start by solving the problem with a single cut in the case where the cake is an orthogonal triangle, then reduce the problem to this case.

**65** There are many ways to proceed, none are easy to find. One possibility is to build the sharing method by induction on $n$. One can begin by thinking of the case in which $n = 3$ and try to reduce to the previous case in which there are only two people.

**66** Consider the pair of soldiers who are the closest to each other.

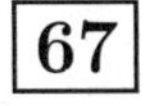
**67** One must consider all possible groups of six pirates.

**68** Each weighing has three possible outcomes; what is the maximum number of weights that one could discriminate between? One will be able to use the method of comparing outcomes to hypotheses, which is detailed in the reminder on the pigeonhole principle (reminder 1).

**69** The answer is 533 bananas. The hardest part is proving one can't do better.

**70** Color some of the small cubes of the cube.

**71** One can begin by assuming that the light is initially off and that the students know this.

**72** The events "student $n$ finds their number" are not necessarily independent: students may find a strategy that links these events. This riddle requires the use of permutation (see reminder 4).

**73** One must use modular arithmetic. If necessary, refer to the reminder on congruence (reminder 6).

**74** One must use the sign (see reminder 4).

**75** It suffices to pour the contents of one pitcher into the other and start this process over as many times as necessary. It remains to prove this result.

**76** Start by reasoning over two days ($q = 2$) by asking yourself, with a fixed number of rats, how many bottles can one test. One must use the method for comparing outcomes to hypotheses (see reminder 1).

 One can use a bijection $\sigma : \mathbb{N} \to \mathbb{Z} \times \mathbb{Z}$ (see reminder 4).

 Think of graph theory.

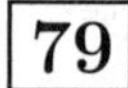 Think of graph theory.

**80** One must use modular arithmetic. If necessary, refer to the reminder on congruence (reminder 6).

**81** One must color the tiles of the bathroom column-wise.

**82** One must color the tiles of the bathroom with three colors.

**83** Start with two houses, then three houses, then four houses, and so on by numbering the houses in base 2 (see reminder 7).

**84** Find the disk in which the swimmer has faster angular speed than the dog. What is the best way to exit this disk?

**85** A natural strategy consists in fixing a sum which we consider sufficient, for example \$100, and cashing the first check if it is greater than \$100, cash the other if not. Does this strategy allow one to do better than $\frac{M_1+M_2}{2}$ in all cases? If not, how to improve it?

**86** Remember the bias of equiprobability: just because there are two outcomes doesn't mean that they are equally likely.

**87** One could translate the conditions of the weights into a matrix and compute the determinant.

**88** One can compare $t_1$ and $t_2$ without computation, but one must be wary of erroneous reasoning: indeed, $t_1 \neq t_2$. What happens when the monkey has typed “abracadabr”?

**89** One is easily convinced by some drawings that the result is true, but to prove it turns out to be quite delicate. There are several possible proofs, more or less advanced. A clever way to proceed is to consider an adapted black and white tiling, like in the bathroom riddle.

**90** Start by connecting points 2 and 3 together; points 4, 5, and 6 together; and points 7, 8, 9, and 10 together.

**91** Start by dividing the classroom into two arbitrary groups. If a student has more friends in one group than the other, move that student over to the other group. What happens?

**92** Bob is tempted to wait a few trials hoping that there will be more revealed black cards than reds. One must show that this strategy (like any other) doesn't change anything, and that Bob will always have a 50/50 chance of getting his prediction right. To do so, one must start by formalizing correctly what the strategy is.

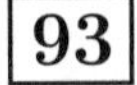

**93** One must improve the following position:

**94** Consider writing the results of each month as a $5 \times 8$ rectangular table.

**95** Number squares in base 2 (see reminder 7).

**96** Equip real sequences with the following equivalence relation: two sequences are equivalent if they are equal after a certain rank (see reminder 3). At some point, the axiom of choice will need to be used.

**97** Use the same equivalence relation as before (see reminder 3).

**98** Still using the same equivalence relation (see reminder 3), but more subtle. Solving the following simpler problem may help: with only 100 students and 100 drawers with integers, each student doesn't have to guess but only give an upper bound for an integer of a closed drawer, still with one mistake allowed.

**99** One could consider all the lines that pass through at least two points of the set, then reason by contradiction with the point that is closest to the lines it isn't on.

**100** One cannot go beyond the fourth line. Finding a configuration to reach the fourth line is in no way obvious, and showing that one cannot reach the fifth line is fiendishly difficult. A shrewd way to proceed is to assign to each square a power of $\phi$, where $\phi$ is the golden ratio minus 1 and thus verifies $\phi^2 + \phi = 1$, as below:

| $\phi^3$ | $\phi^2$ | $\phi^1$ | 1 | $\phi^1$ | $\phi^2$ | $\phi^3$ |
|---|---|---|---|---|---|---|
| $\phi^4$ | $\phi^3$ | $\phi^2$ | $\phi^1$ | $\phi^2$ | $\phi^3$ | $\phi^4$ |
| $\phi^5$ | $\phi^4$ | $\phi^3$ | $\phi^2$ | $\phi^3$ | $\phi^4$ | $\phi^5$ |
| $\phi^6$ | $\phi^5$ | $\phi^4$ | $\phi^3$ | $\phi^4$ | $\phi^5$ | $\phi^6$ |
| $\phi^7$ | $\phi^6$ | $\phi^5$ | $\phi^4$ | $\phi^5$ | $\phi^6$ | $\phi^7$ |
| $\phi^8$ | $\phi^7$ | $\phi^6$ | $\phi^5$ | $\phi^6$ | $\phi^7$ | $\phi^8$ |
| $\phi^9$ | $\phi^8$ | $\phi^7$ | $\phi^6$ | $\phi^7$ | $\phi^8$ | $\phi^9$ |

We call the sum of the values assigned to the squares where there are marbles the energy of a configuration. One can show that the energy cannot decrease and deduce from it that it is impossible to reach the square with energy one.

# Part III
# Solutions

**1** It is sufficient for the blacksmith to open three links on a single chain, then use them to link all other chains together.

**2** It takes 1 minute for an ant to fall in the worst case. One can convince oneself easily that two ants also take 1 minute to fall in the worst case and that it is the same for three ants. The answer is always the same: 1 minute. It is sufficient to notice that when two ants bump into each other, everything happens as if they simply passed through one another. Thus, it doesn't matter whether there is one ant or 50.

**3** The third friend paid \$8; since each person has contributed equally, the total money spent is $8 \times 3 = \$24$. Each cake is thus worth $\frac{24}{8} = \$3$. The first friend made five cakes, so she spent $5 \times 3 = \$15$ and she must be given back $15 - 8 = \$7$. The second friend made three cakes, worth \$9 total, so she must be given back \$1. In the end, the third friend must give \$7 to the first and \$1 to the second.

**4** The mass of ketchup on the plate represents 99% of the total mass, which is 1 kg. We conclude that the fry weighs 10 grams and that there are 990 grams of ketchup. One would think that since 990 grams of ketchup represents 99% of the total mass, then 980 grams would represent 98% of the total mass and it suffices to remove 10 grams of ketchup. This is wrong since 980 grams is not 98% of the total mass: $\frac{980}{980+10} \simeq 98.99\% \neq 98\%$. Once the excess ketchup has been removed, the fry, which still weighs 10 grams, will represent 2% of the total mass. The total mass is then 500 grams, since $\frac{10}{500} = 2\%$ and there is 490 grams of ketchup left. Thus, the diner must remove 500 grams of ketchup!

**5** By walking 1,000 km south, the explorer will find himself on a parallel. He will then, by going west, walk along this parallel for 1,000 km before going back up north. If he did not start from the North Pole, he must necessarily return to the point where he turned onto the parallel before going back north. For example, if this parallel has a perimeter of 1,000 km, it works: any point 1,000 km north of this parallel is a possible starting point, since the explorer will descend onto this parallel, loop around it, then go back up north, like in the figure below (not to scale).

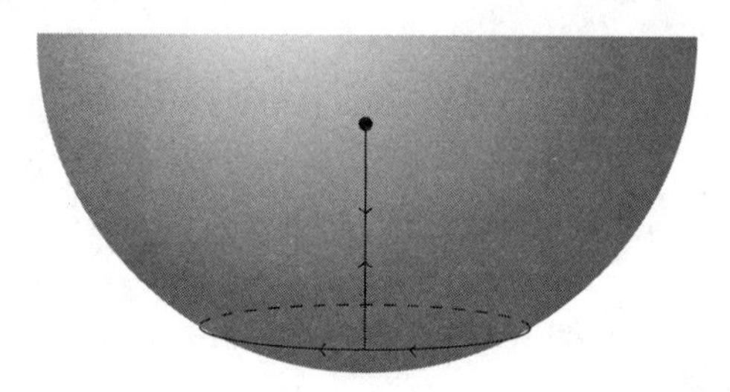

Is this the only possibility? No! The parallel can also have a perimeter of $\frac{1{,}000\,\text{km}}{2}$, $\frac{1{,}000\,\text{km}}{3}$, $\frac{1{,}000\,\text{km}}{4}$, and so forth. We thus have all the possible start points:

- The North Pole (the exception)
- The parallels lying 1,000 km north of the parallels whose perimeter is $\frac{1{,}000\,\text{km}}{n}$ for some nonzero integer $n$.

**6** It suffices to light the first vine at both ends and to simultaneously light the second at one end. The first vine will burn for 2 minutes. As soon as it is finished, set the second end of the second vine on fire, which will then burn for 1 more minute. Two minutes and 1 minute keeps the desired time.

**7** The patient takes an A pill out of the box, cuts it in two, and swallows half. He then halves the three pills he has in hand and swallow half of each one. The next day he swallows the remaining four halves.

**8** One is tempted to answer 60 km/h since 40 km/h is the average of 20 km/h and 60 km/h. Let's see why this is wrong. Let $d$ be the distance between the house and the university. Going 20 km/h, it takes the student a time $t_1$ to go to the university, $20\,\text{km/h} = \frac{d}{t_1}$ thus $t_1 = \frac{d}{20\,\text{km/h}}$. The return trip takes him $t_2 = \frac{d}{60\,\text{km/h}}$. The total time of the round trip is $t_1 + t_2$, for a total distance of $2d$; his average speed is thus

$$v = \frac{2d}{t_1 + t_2} = \frac{2}{\frac{1}{20\,\text{km/h}} + \frac{1}{60\,\text{km/h}}} = 30\,\text{km/h (and not } 40\,\text{km/h).}$$

In fact, there is no solution to this problem. The only way for the student to double his average speed is to have an infinite return speed, that is, to teleport! Indeed, if he returns instantly, letting $t$ be the time of the outbound leg, the total time for the round trip is $t$, and the average speed of the round trip is thus $v = \frac{2d}{t} = 2 \times 20\,\text{km/h} = 40\,\text{km/h}$.

In reality, since he does not teleport, his average speed on the round trip will always be strictly less than 40 km/h.

**9** The first challenge is in posing the problem properly. Let $v$ be the speed of the boat absent any current; when the speed of the current is $v_c$, the speed of the boat is $v + v_c$ if it is going downstream and $v - v_c$ if not. Then, one must absolutely not average the two speeds and conclude that the boat will take the same amount of time whatever the speed of the current. This reasoning is intuitive, but it is wrong. Like we saw in the previous riddle, the average speed $v_m$ on the round trip when there is a current $v_c < v$ (otherwise the boat can't come back at all) is

$$v_m = \frac{2}{\frac{1}{v-v_c} + \frac{1}{v+v_c}} \quad \text{thus} \quad v_m = \frac{v^2 - v_c^2}{v}.$$

$v_m$ is a decreasing function of $v_c$. If there is more current than usual, the average speed will be lower and thus the total time will be greater than usual.

One can recover this result quickly, but without really proving it, by reasoning about the limit case. If the speed of the current is equal to the speed of the boat, the boat will be twice as fast going downstream but will never be able to return since it will have a speed of 0. The total travel time will thus be infinite.

**10** The proportion of boys will not change. The following false reasonings are often heard.

– There will be more boys since a family that wants a boy gets two tries while a family wanting a girl must get a girl on the first try.

– There will be more girls since the last child born has a 50/50 chance of being a girl, whereas the first child born is necessarily a girl.

The best way to reason is to consider the point of view of a midwife. Each child she sees be born has a 50/50 chance of being a boy, so the overall proportion of girls to boys in the country isn't changed.

**11** Neither. The amounts of each are equal and there is no need for computation to show it! The total amounts of coffee and milk remain unchanged during the operation; they stay equal to 200 mL. Suppose, for example, there is 197 mL of coffee and 3 mL of milk in the coffee cup, there is $200 - 3 = 197$ mL of milk and $200 - 197 = 3$ mL of coffee in the milk cup.

**12** The coffee lover will never succeed in obtaining a cup containing as much coffee as milk. Suppose for contradiction that he could obtain a 50/50 mix. Let's focus on the last operation that made the mix balanced. The coffee lover puts a spoonful of the first cup into the second then one spoonful of the second cup into the first. He then obtains a 50/50 mix. The spoon that went from the first cup to the second thus came from a 50/50 mix cup, and it turned the second cup into a 50/50 mix. Necessarily, the second cup thus contained a 50/50 mix. This is a contradiction since we assumed that before this final step wasn't an even 50/50 mix.

**13** It is sufficient to ask either one of the guardians "What would the other guardian say if I asked them if this is the door to hell?" If the guardian you ask is the liar, they will answer "yes" if it is the door to heaven, and "no" if it is the door to hell. If the guardian you ask tells the truth, they will answer "yes" if it is the door to heaven and "no" if it is the door to hell as that is what the liar will say. Thus, in both cases, the guardian will answer "yes" if the door leads to heaven and "no" otherwise.

**14** For an integer $k$, let $n_k$ be the number of ways to climb a staircase with $k$ steps. We have $n_1 = 1$, $n_2 = 2$, $n_3 = 3$, and so on, and we seek

$n_{10}$. Let $k$ be between 3 and 10. When the kangaroo climbs a staircase of $k$ steps, it can choose to climb the first step, in which case it has a $k-1$ step staircase left to climb, or to jump the first step, in which case it has a $k-2$ step staircase to climb. We thus obtain the following relation:

$$n_k = n_{k-1} + n_{k-2} \, .$$

One recognizes the famous Fibonacci sequence, in which each new term is the sum of the two previous terms. The first few terms are:

$$1,\ 2,\ 3,\ 5,\ 8,\ 13,\ 21,\ 34,\ 55,\ 89.$$

Thus, $n_{10} = 89$. We can also show that:

$$n_k = \frac{1}{\sqrt{5}} \left( \left( \frac{1+\sqrt{5}}{2} \right)^{k+1} - \left( \frac{1-\sqrt{5}}{2} \right)^{k+1} \right)$$

and find the value of $n_{10}$ this way.

**15** The trains start out 400 km from each other and both move at a speed of 200 km/h, so they drive for 1 hour before passing each other; thus, the fly flies for 1 hour. Since it flies at 400 km/h it covers in total exactly 400 km during this hour. One could also compute an infinite sum, but it takes longer to get the same result.

**16** The correct question to ponder is at what time did the boy encounter his mother on the road? His mother traveled 10 minutes less than usual, so she turned around 5 minutes earlier than usual; this means the mother met her son at 15:55 (3:55 PM) instead of 16:00 (4 PM). The little boy thus walked for 55 minutes.

**17** Consider the small middle cube: one must necessarily perform six cuts (one per face) to separate it from the other small cubes. One can rearrange the pieces in any way one wishes, but one cannot accomplish the task in fewer than these six cuts.

**18** Notice first of all that it is sufficient to bring the sum to 89. Indeed, the second player, who must give an integer between 1 and 10, will only be able to move the sum to between 90 and 99, which allows the first player to reach 100 and win the game.

Likewise, if the first player brings the sum to 78, it is over since it is possible to bring the sum to 89, and from there, to 100. One can go down by 11 each time: the winning sums (when it's the opponent's turn to play) are 89, 78, 67, 56, 45, 34, 23, 12, 1. Thus, it suffices that the first player start by playing a 1, then bring the sum to 12, then 23, and so on until 89 and finally to 100.

**19** The trick is to compute how many living rooms the three painters can paint together in $2 \times 3 \times 5 = 30$ hours, or how many times they can paint the same room in these 30 hours, which is the same question. In 30 hours, the first painter will paint the living room $\frac{30}{2} = 15$ times, the second $\frac{30}{3} = 10$ times, and the third $\frac{30}{5} = 6$ times. In total, they will paint the room $15 + 10 + 6 = 31$ times in 30 hours, so it will take them $\frac{30}{31}$ hours to paint the room once, which is roughly 58 minutes.

**20** Let us reason by contradiction and assume that each guest shook a different number of hands. Note $n$ as the number of guests. The number of hands shaken for each guest is between 0 and $n-1$, which is $n$ possibilities. Since we supposed that each guest has shaken a different number of hands, there is a guest who shook 0 hands, one who shook 1 hand, one who shook 2 hands, ..., and a guest who shook $n-1$ hands. But if a guest shook $n-1$, then he shook everyone else's hands, which is not possible if one guest shook 0 hands: a contradiction. Therefore, there must be two people who shook the same number of hands.

**21** One can easily convince oneself of the result graphically by drawing the distance of the executive to her house as a function of time, as shown in the following figure:

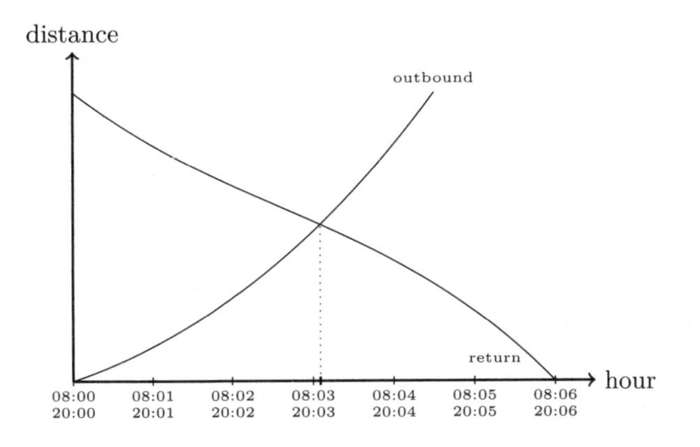


It is clear that there is at least one point where the two curves intersect, and the time associated to this point solves the problem. In the depicted example, the executive passes at the same place as 12 h earlier at 20:03 (8:03 PM) plus a few seconds.

For a perfectly rigorous proof, one can use the intermediate value theorem to prove the existence of the intersection point.

**22** It is sufficient to place the first coin right in the center of the table and then play symmetrically (central symmetry) to the employer. If a spot is free for the employer, then the symmetric spot is free; thus, the student cannot lose. Since the table is finite, he will eventually win.

**23** Let $p$ be the probability that two students have their birthday on the same day. The easiest way to compute $p$ is to compute $1-p$, the probability that all students have different birthdays. To do so, we will count the number of cases in which all students have their birthdays on different days, then divide it by the total number of cases. The total number of cases is $365^{23}$. The number of cases in which the students have different birthdays is $365 \times (365-1) \times \cdots \times (365-22)$. Indeed, there are 365 possibilities for the birthday of the first student, $365-1$ for the second since it must differ from the first, and so on. We thus have

$$1-p = \frac{365 \times (365-1) \times \cdots \times (365-22)}{365^{23}}$$

$$p = 1 - \frac{365!}{(365-23)! \times 365^{23}}.$$

Using a computer, $p \simeq 0.51$. Despite expectations, there is a greater than 50/50 chance that two students have their birthday on the same day!

For $n$ students, with $n \leq 365$, the probability that two students have their birthday on the same day is

$$p_n = 1 - \frac{365!}{(365-n)! \times 365^n}.$$

We note that $p_n$ goes to 1 quickly. For example, $p_{50} \simeq 0.97$ and $p_{80} \simeq 0.9999$.

**24**

- With one snake: The snake eats the mouse in peace.
- With two snakes: Each snake knows that if it eats the mouse, it could get eaten by the other snake (which is safe since it would be alone). Thus, nothing happens, the two snakes remain locked staring at each other unable to eat the mouse.
- With three snakes: We saw that with two snakes, nothing would happen. The fastest snake thus dives on the mouse and leaves the other two to stare at each other. The fastest snake knows he won't be eaten.

This reasoning easily extends: with four snakes nothing happens, since if one snake eats the mouse we return to the three snake case and it gets eaten. With five snakes, the fastest eats the mouse, with six snakes nothing happens, and so on. We conclude by induction: with fifteen snakes, the fastest eats the mouse.

**25** The problem is the computation of a conditional probability: the probability that a person is sick given that the test is positive. We will thus use Bayes' rule (see reminder 8). Let us define some notations. $S$ is for sick, $H$ is for healthy, and $P$ is for positive. Let $\mathbb{P}()$ denote probabilities (see reminder 8); by Bayes' rule

$$\mathbb{P}(S|P) = \frac{\mathbb{P}(P|S) \times \mathbb{P}(S)}{\mathbb{P}(P)}.$$

Since each person is either $S$ or $H$, $\mathbb{P}(P) = \mathbb{P}(P \text{ and } S) + \mathbb{P}(P \text{ and } H)$

$$\mathbb{P}(S|P) = \frac{\mathbb{P}(P|S) \times \mathbb{P}(S)}{\mathbb{P}(P|S) \times \mathbb{P}(S) + \mathbb{P}(P|H) \times \mathbb{P}(H)}$$

$$\mathbb{P}(S|P) = \frac{0.99 \times 0.001}{0.99 \times 0.001 + 0.01 \times (1 - 0.001)}$$

We obtain $\mathbb{P}(S|P) \simeq 9\%$. Despite the apparent accuracy of the test, a person with a positive screening test has only a 1-in-10 chance of being sick!

**26** Here are the complete results of shots taken in both matches:

| | Team A | Team B |
|---|---|---|
| Match 1 | 93% (81/87) | 87% (234/270) |
| Match 2 | 73% (192/263) | 69% (55/80) |
| Total | 78% (273/350) | 83% (289/350) |

Team A made 93% of shots in the first match and 73% in the second. The team made 78% of the total number of shots they took throughout the game. Team B had lower success rates in both the first and the second matches, but overall had a higher success rate!

This phenomenon can be explained in the following manner: an average percentage is not computed as the average of the percentages; it depends on some weighting. For example, the total success rate of team A is not computed by averaging 93% and 73%:

$$\frac{93\% + 73\%}{2} = 83\% \neq 78\%.$$

One must take into account the fact that the 73% has more weight than the 93% because it is computed with more taken shots. Let $N_1^A$ and $N_2^A$ be the number of shots taken by team A in the first and second matches respectively (here $N_1^A = 87$ and $N_2^A = 263$). The average percentage is computed in the following way:

$$\frac{N_1^A \times 93\% + N_2^A \times 73\%}{N_1^A + N_2^A} = \frac{81 + 192}{87 + 263} = 78\%.$$

Thus, the average percentage is situated somewhere between the two percentages but not necessarily in the middle. It can thus be very close to one of the two ends if the number of shots taken into account for each of the matches is very different. The average percentage of team A is between 93% and 73%, but with more weight on the 73%, it ends up at 78%. The average percentage of team B is between 87% and 69%, but with more weight on the 87%, it ends up at 83%.

The numbers for this example are from a medical study conducted to test the performance of two treatments A and B (team A and team B) against kidney stones.[2]

Here is another example to better understand the phenomenon. Consider four urns, numbered from 1 to 4. Each urn contains black balls and white balls. Urns 1 and 3 are of type $A$, while 2 and 4 are of type $B$. There is a higher percentage of black balls in urn 1 than in urn 2 (100% versus 99%). Also, there is a higher percentage of black balls in urn 3 than in urn 4 (1% versus 0%). Consider urns 1 and 2 as one set, and urns 3 and 4 as another set. Within each set, the percentage of black balls is always higher in the type $A$ urn than in the type $B$ urn. However, overall, the percentage of black balls is higher in urns of type $B$ (urn 2 and urn 4) than in urns of type $A$ (urn 1 and urn 3).

| | Type A | Type B |
|---|---|---|
| Urns 1 and 2 | 100% (1/1) | 99% (99/100) |
| Urns 3 and 4 | 1% (1/100) | 0% (0/1) |
| Total | 2% (2/101) | 98% (99/101) |

[2] C. R. Charig, "Comparison of treatment of renal calculi by open surgery, percutaneous nephrolithotomy, and extracorporeal shockwave lithotripsy", *British Medical Journal, Clinical Research Ed.* 292, no. 6524 (March 1986): 879–882.

**27** With each throw, the basketball player increases his success rate if he scores and decreases it otherwise. After the first throw, his success is 0% but it will increase up to 80%. At some point, say between the $(n-1)$th and the $n$th throw, for some $n > 1$, the player's rate will go from being strictly less than 75% to being strictly greater than 75%. Let $p$ be the number of successful throws by the $n$th throw. At the previous throw, the player's success rate was lower, so he had only $p-1$ successes. Thus

$$\frac{p-1}{n-1} < \frac{3}{4} = 75\% \text{ and } \frac{p}{n} \geq \frac{3}{4} = 75\%$$

This implies

$$4(p-1) < 3(n-1) \text{ and } 4p \geq 3n$$

We thus have

$$0 \leq 4p - 3n < 1.$$

But $4p - 3n$ is a whole number, so $4p - 3n = 0$, and, at the $n$th throw, the basketball player has a success rate of exactly 75%.

There is another proof which is more visual: one draws the sequence defined by recurrence starting from 0 at 0 and increasing by 1 when the basketball player succeeds at a throw and decreasing by 3 otherwise. For example, we obtain the graph below if the basketball player failed the first and fourth throws.

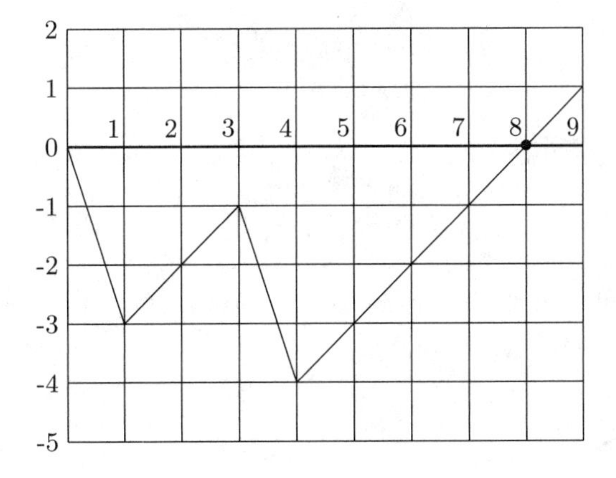

At the end, the sequence must be positive because the player's success rate is higher than 75%. When the sequence cancels (at 8 here), the success rate is 75%. Indeed, there are three times more successful throws than failed ones.

**28** Let us proceed in order. To start, we know that the product of the ages is 36. Let us write all possible products:

$$1 \times 1 \times 36 = 36$$
$$1 \times 2 \times 18 = 36$$
$$1 \times 3 \times 12 = 36$$
$$1 \times 4 \times 9 = 36$$
$$1 \times 6 \times 6 = 36$$
$$2 \times 2 \times 9 = 36$$
$$2 \times 3 \times 6 = 36$$
$$3 \times 3 \times 4 = 36$$

For each product, let us compute the sum of ages:

$$1 \times 1 \times 36 = 36 \quad 1 + 1 + 36 = 38$$
$$1 \times 2 \times 18 = 36 \quad 1 + 2 + 18 = 21$$
$$1 \times 3 \times 12 = 36 \quad 1 + 3 + 12 = 16$$
$$1 \times 4 \times 9 = 36 \quad 1 + 4 + 9 = 14$$
$$1 \times 6 \times 6 = 36 \quad 1 + 6 + 6 = 13$$
$$2 \times 2 \times 9 = 36 \quad 2 + 2 + 9 = 13$$
$$2 \times 3 \times 6 = 36 \quad 2 + 3 + 6 = 11$$
$$3 \times 3 \times 4 = 36 \quad 3 + 3 + 4 = 10$$

We observe that there are only two products which yield the same sum. Recall that the postman couldn't conclude knowing the sum of ages. Therefore the number of the house across the street is 13, and the postman hesitated between $(1, 6, 6)$ and $(2, 2, 9)$. There is an eldest, therefore the children are 2, 2, and 9.

**29** One possible solution:

For readers who want to know where this solution comes from, let us use the notations used in the hint of this puzzle (see the section on Hints). If we consider the sequence $aba^{-1}b^{-1}$ (also known as the commutator of $a, b$ and written $[a, b]$), we notice that upon removing the first nail, we drop all occurrences of $a$ and $a^{-1}$ in the sequence so that it becomes $bb^{-1}$. This means the rope turns clockwise and then counterclockwise around the second nail, and the painting falls. The same thing occurs if the second nail is removed.

This corresponds to the solution given above. This approach is useful to understand the solution, but also to generalize it: with three nails, we add the symbols $c, c^{-1}$, and see that $[[a, b], c] = aba^{-1}b^{-1}cbab^{-1}a^{-1}c^{-1}$ is a solution. Indeed, remove any nail, (e.g., drop all $a$ and $a^{-1}$), then the expression simplifies, meaning that the painting falls. For a more detailed explanation and broader generalization, we refer to the article: E. D. Demaine, et al. (2014), "Picture-hanging puzzles," *Theory of Computing Systems*, 54(4): 531–550.

**30** Strange question! One could indeed think that nothing will happen since everyone already knows that there is at least one unfaithful spouse. Each cheated spouse knows 49 others, all the other inhabitants know 50. However, something will happen; let us see why.

Let us imagine there were only two unfaithful spouses instead of 50. Let us step into the shoes of a cheated spouse: I am being cheated on, but I don't know it. I only know one cheated spouse. When I discover the message on the wall, I think to myself that the only cheated spouse I know will learn that they are being cheated by reading the message and break up with their spouse that very evening. Bad luck, I realize the next morning that no one has broken up. I conclude that there isn't just one cheated spouse but two: the one I know of and...myself! Confident in this logical reasoning, I separate from my spouse that evening. The other cheated spouse follows the same reasoning and acts the same way. Thus, when there are two unfaithful spouses and thus two cheated ones, both couples break up on the second evening.

What happens now if there are three unfaithful spouses? Once again, we step into a cheated spouse's shoes. I know two people being cheated on and I think they will separate from their spouses on the second evening. Yet nothing happens on the second evening, so I break up our couple on the third evening. We show in this way, by recurrence, that nothing happens for the first 49 days, but on the 50th evening, all the cheated spouses break up at the same time!

**31** The probability, for a couple with at least one daughter to also have a boy is $\frac{2}{3}$. Indeed, couples with two children are divided into four equally likely categories (denoting G for girl and B for boy):

| | |
|---|---|
| GG (G then G) | GB (G then B) |
| BG (B then G) | BB (B then B) |

Couples with at least one daughter are those which are GG, BG, or GB. Among these couples, two thirds (those with BG or GB) have at least one boy: the probability we wanted is indeed $\frac{2}{3}$.

We now ask the couple if they have a daughter called Sophie. If they answer yes, the probability that they have a boy is the probability they have a boy given they have a daughter called Sophie (see reminder 8). To compute this conditional probability, let us represent, in gray, those couples who have at least one child called Sophie (these couples are from the population of couples with two children that was mentioned earlier). There is a small proportion in the square GB, as many in the square BG, and about twice as many in the square GG.

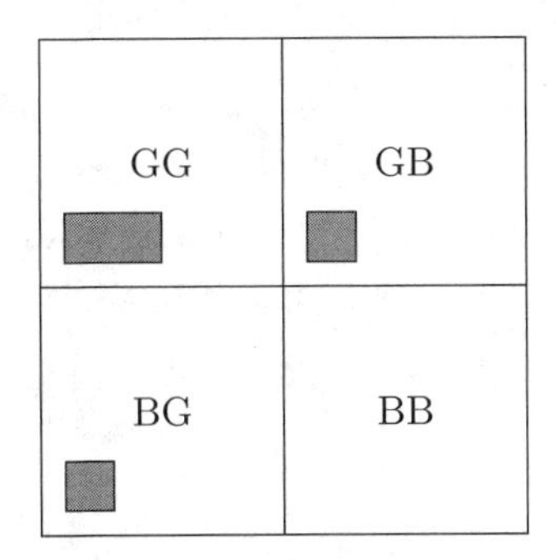


Since there is as much gray in GG than in both other boxes combined, (GB and BG), knowing that there is a daughter called Sophie, we can calculate that there is equal probability that there are two girls or one girl and one boy. The probability we wanted is thus about $\frac{1}{2}$. Let us compute it more precisely. Let us note $S$ as the event "at least one child is called Sophie" and $B$ as the event "at least one child is a boy." Let us begin by computing $\mathbb{P}\left(\mathrm{GG}|\mathrm{S}\right)$.

$$\mathbb{P}\left(\mathrm{GG}|\mathrm{S}\right) = \frac{\mathbb{P}\left(\mathrm{S}|\mathrm{GG}\right) \times \mathbb{P}\left(\mathrm{GG}\right)}{\mathbb{P}\left(\mathrm{S}\right)}$$

$$\mathbb{P}\left(\mathrm{GG}|\mathrm{S}\right) = \frac{\left(1 - \mathbb{P}\left(\text{not S}|\mathrm{GG}\right)\right) \times \mathbb{P}\left(\mathrm{GG}\right)}{\mathbb{P}\left(\mathrm{S}|\mathrm{GG}\right) \times \mathbb{P}\left(\mathrm{GG}\right) + \mathbb{P}\left(\mathrm{S}|\text{GB or BG}\right) \times \mathbb{P}\left(\text{GB or BG}\right)}$$

$$\mathbb{P}\left(\mathrm{GG}|\mathrm{S}\right) = \frac{\left(1-(1-\epsilon)^2\right) \times \frac{1}{4}}{\left(1-(1-\epsilon)^2\right) \times \frac{1}{4} + \frac{\epsilon}{2}} = \frac{2\epsilon - \epsilon^2}{4\epsilon - \epsilon^2}$$

Since $\mathbb{P}\left(\mathrm{B}|\mathrm{S}\right) = 1 - \mathbb{P}\left(\mathrm{GG}|\mathrm{S}\right)$,

$$\mathbb{P}\left(\mathrm{B}|\mathrm{S}\right) = \frac{1}{2 - \frac{\epsilon}{2}}$$

which is indeed approximately equal to $\frac{1}{2}$ for small $\epsilon$. It is reassuring to notice that in the case in which all girls are called Sophie, which is the case for $\epsilon = 1$, we recover $\frac{2}{3}$.

Let us now study the third case. The couple answered yes when asked if they had at least one daughter. We conclude that they have a probability of $\frac{2}{3}$ of having a boy. They now add that they also have at least one daughter called Sophie. Now that we know they have a daughter called Sophie, we would be tempted to say that the probability that the couple has a boy is no longer $\frac{2}{3}$ but $\mathbb{P}(\mathrm{B}|\mathrm{S})$. This reasoning applies to whatever name is announced by the parents, since we assume all names have the same probability of being given at birth. For example, if the couple announces they have at least one child called Laura instead, the probability also goes from $\frac{2}{3}$ to $\mathbb{P}(\mathrm{B}|\mathrm{Laura}) = \mathbb{P}(\mathrm{B}|\mathrm{S})$. Thus, regardless of what the parents say, the probability goes from $\frac{2}{3}$ to $\mathbb{P}(\mathrm{B}|\mathrm{S})$, which is absurd! In fact, knowing that a child is called Sophie isn't sufficient to conclude that the probability that the couple has a boy is $\mathbb{P}(\mathrm{B}|\mathrm{S})$; it all depends on how this information was obtained.

Let us take a simple example to convince ourselves. Suppose there are three cards bearing respectively the letters a, b, and c, and lying face down on a table. The game consists of choosing one of the three face-down cards. Let $A$ be the event "the chosen card is card a"; we have obviously $\mathbb{P}(A) = \frac{1}{3}$. Group 1 is made up of the cards a and b and group 2 is made up of cards a and c. Note $\mathrm{G}_1$ (respectively $\mathrm{G}_2$) as the event "the card chosen belongs to group 1 (respectively 2)." We thus have

$$\mathbb{P}(A|\mathrm{G}_1) = \mathbb{P}(A|\mathrm{G}_2) = \frac{1}{2}$$

We now choose a card face down and set it aside. We know the chance of it being card a is $\frac{1}{3}$. If we were to ask a friend to look at the card and tell us a group to which it belongs, then regardless of the reply (group 1 or group 2), if we were to trust the conditional probabilities, the probability that the chosen card is card a would systematically jump from $\frac{1}{3}$ to $\frac{1}{2}$, which is absurd! Thus, it is not sufficient to be told "the chosen card is in group 1" for us to be able to conclude that the probability that it is card a jumps from $\mathbb{P}(\mathrm{A})$ to $\mathbb{P}(\mathrm{A}|\mathrm{G}_1)$. The problem comes from the fact that we obtain the information "by force." When card a is drawn, the decision of the friend is what matters: does he always answer group 1, or does he draw at random between group 1 and group 2? If he draws at random, then half of the cards will end up in "his" group 1, which means if he answers "group 1," the card he is looking at is twice as likely to be card c than card a. This is not the same as drawing a card at random from group 1, where cards a and c are equiprobable. He does not answer "group 1" if and only if he holds card a or c, since for half of the a cards he responds with "group 2": "his" group 1 is not representative of nor equivalent to the real group.

Instead of asking the friend to give us a group to which the card belongs, if we directly ask him if it is in group 1, when he answers "yes" the probability does jump from $\mathbb{P}(\mathrm{A}) = \frac{1}{3}$ to $\mathbb{P}(\mathrm{A}|\mathrm{G}_1) = \frac{1}{2}$, since he responds "yes" if and only if the card belongs to group 1.

The case of first names is much the same. When the couple has two daughters, they must choose a name to announce (same as when the friend picks a group for card a). Thus, if we assume that when they are asked to give the name of (one of) their girl(s), their answer is based on a coin toss. If they have two girls, the probability that this couple has a boy stays $\frac{2}{3}$. On the other hand, if we assume that a couple always tries to announce Sophie if possible, the probability will indeed go from $\frac{2}{3}$ to $\mathbb{P}(\mathrm{B}|\mathrm{S})$. Therefore, if we ask a large number of couples, those who answer Sophie to the second question are exactly those who have at least one child called Sophie, of whom a proportion of $\mathbb{P}(\mathrm{B}|\mathrm{S})$ have a boy.

**32** At first glance, there is no reason to change. Nevertheless, by choosing a door that doesn't contain a car, the host gives an indication of the car's position.

Let us number the doors from 1 to 3, and suppose that door 1 was chosen. The car can be located in one of three possible places: behind door 1, behind door 2, or behind door 3.

– If the car is behind door 1, one must of course not change doors.

– If the car is behind door 2, the host opens door 3, and door 2 is the only option left, then one must change doors.

– If the car is behind door 3, the host opens door 2, and door 3 is the only option left, then one must change doors.

On average, changing doors allows the contestant to win the car two out of three times, so she must change doors. If the contestant plays too quickly and fails to change her door, she will have deprived herself of the chance to double her chance of winning.

One can also show this result using Bayes' rule (see reminder 8). Let $C$ (for car) denote the event "choosing the car door at the start" and let $\overline{C}$ be its complement: "not choosing the car door at the start." Let $W$ (for win) denote the event "choosing the car door at the end." We seek the probability of wining the car $\mathbb{P}(W)$.

With the strategy of sticking to the same door, $\mathbb{P}(W) = \frac{1}{3}$.

With the strategy of changing doors and conditioning with respect to events $C$ and $\overline{C}$

$$\mathbb{P}(W) = \mathbb{P}(W|C) \times \mathbb{P}(C) + \mathbb{P}\left(W|\overline{C}\right) \times \mathbb{P}\left(\overline{C}\right)$$

But, $\mathbb{P}(W|C) = 0$ since if one had chosen the right door from the start, then changing is a guaranteed loss. On the other hand, if one hadn't chosen the right door from the start, it is a guaranteed win; thus, $\mathbb{P}\left(W|\overline{C}\right) = 1$. In the end we have

$$\mathbb{P}(W) = 0 \times \frac{1}{3} + 1 \times \frac{2}{3} = \frac{2}{3}$$

By contrast, when the host opens a door behind which there is no car by pure chance, without purposefully keeping the suspense up, changing doors does not increase the likelihood of winning.

To convince oneself, a first approach is to make a tree like the following one:

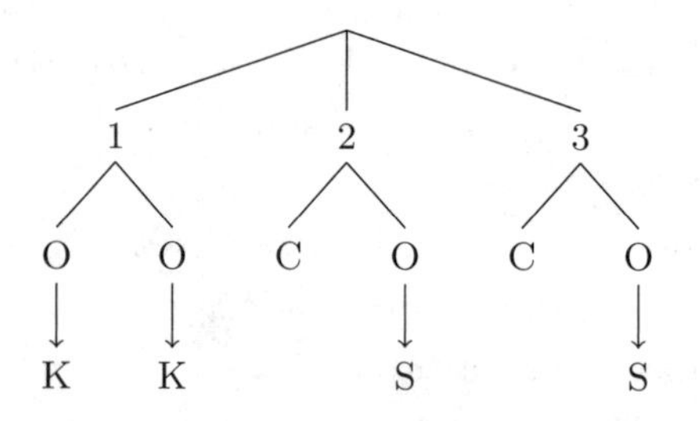


The first line is the number of the door chosen, number 1 being the one behind which the car is hidden. The second line corresponds to what the host sees: $O$ for other, $C$ for car. If $C$ is revealed, then there is nothing to ask the contestant; otherwise, the third line shows the choice to make in consequence: $K$ to keep and $S$ to switch doors. Since there are as many $K$ choices as $S$ choices, we can say the strategies are equivalent. However, this reasoning must be made rigorous, and for this, we will apply Bayes' rule.

Let $H$ (for hidden) be the event "the host doesn't reveal the car," and let $\overline{H}$ be its complement, the event "the host reveals the car." By assuming the host hasn't revealed the car, we no longer compute $\mathbb{P}(W)$ but rather $\mathbb{P}(W|H)$. We have

$$\mathbb{P}(W|H) = \frac{\mathbb{P}(W \text{ and } H)}{\mathbb{P}(H)}$$

We can compute $\mathbb{P}(H)$, even if it is not necessary, by conditioning on events $C$ and $\overline{C}$:

$$\mathbb{P}(H) = \mathbb{P}(H|C) \times \mathbb{P}(C) + \mathbb{P}\left(H|\overline{C}\right) \times \mathbb{P}\left(\overline{C}\right)$$

$\mathbb{P}(H|C) = 1$ (if we chose the car the host will not reveal it) and $\mathbb{P}\left(H|\overline{C}\right) = \frac{1}{2}$ (if we didn't the host has a 50/50 chance of revealing it), so that $\mathbb{P}(H) = 1 \times \frac{1}{3} + \frac{1}{2} \times \frac{2}{3} = \frac{2}{3}$. It remains to compute $\mathbb{P}(W \text{ and } H)$ to know $\mathbb{P}(W|H)$. We have

$$\mathbb{P}(W \text{ and } H) = \mathbb{P}(W \text{ and } H|C) \times \mathbb{P}(C) + \mathbb{P}\left(W \text{ and } H|\overline{C}\right) \times \mathbb{P}\left(\overline{C}\right)$$

With the strategy of sticking to the same door, $\mathbb{P}(W \text{ and } H|C) = 1$ and $\mathbb{P}\left(W \text{ and } H|\overline{C}\right) = 0$, so $\mathbb{P}(W \text{ and } H) = 1 \times \frac{1}{3} + 0 \times \frac{2}{3} = \frac{1}{3}$. Thus, $\mathbb{P}(W|H) = \frac{1}{2}$. With the strategy of changing doors, we have $\mathbb{P}(W \text{ and } H|C) = 0$ and $\mathbb{P}\left(W \text{ and } H|\overline{C}\right) = \frac{1}{2}$, so $\mathbb{P}(W \text{ and } H) = 0 \times \frac{1}{3} + \frac{1}{2} \times \frac{2}{3} = \frac{1}{3}$. Thus, $\mathbb{P}(W|H) = \frac{1}{2}$. Both strategies lead to the same probability of winning.

**33** It is clear that two squares may not form another square (unless one stacks the pieces, but this is forbidden); hence, 2 is an impossible number. Another impossible number is 3. However, 4 is easy: with 4 identical squares, one can make a bigger square. Finally, one is easily convinced that 5 is impossible. All other numbers are possible. The numbers 6, 7, and 8 are possible for a large square:

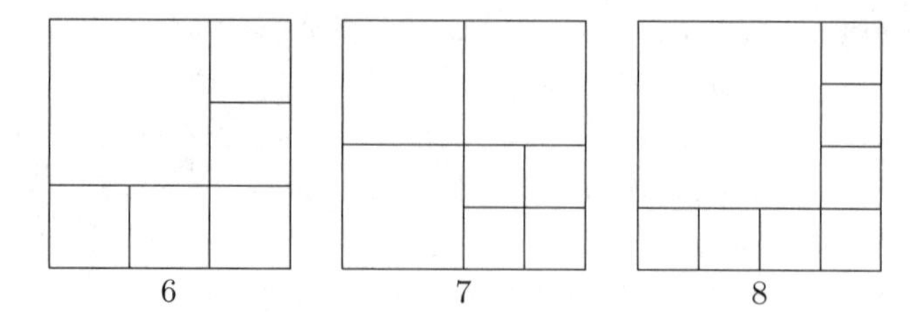


Furthermore, one notices that if one can make $n$, then one can make $n+3$. It suffices to divide any of the squares into four to pass from $n$ to $n+3$. Thus, one can make $9=6+3$, $10=7+3$, $11=8+3$, and so on by threes.

**34** The Good must miss on purpose. To convince ourselves, let us study the case where he shoots at the Ugly, then the case where he shoots at the Bad, and finally where he misses on purpose.

**The Good shoots at the Bad** This strategy is undoubtedly bad, since if he succeeds, he faces off with the Ugly, which leads to his certain death. However, if he misses, he has the same probability of survival than if he missed on purpose. Therefore, his survival probability is strictly lower than the one obtained by missing on purpose.

**The Good shoots at the Ugly** This strategy is the most intuitive: begin by eliminating the most dangerous opponent. Let us calculate $p$, which is the probability that the Good wins if he ends up in a duel against the Bad, with the Bad shooting first. Let $q$ be the probability that the Good wins if it's his turn to shoot against the Bad. Since either the Bad kills the Good or he gives the Good a probability $q$ of winning the duel, $p=\frac{1}{2}q$. Because the Good has a 1-in-3 chance of killing the Bad and a 2-in-3 chance of ending up in the situation where the Bad shoots first, $q=\frac{2}{3}p+\frac{1}{3}$. This gives us $p=\frac{1}{4}$. This strategy is interesting only if $p$ is greater than the probability of winning by shooting into the air, which we must now compute.

**The Good misses his shot on purpose** The Good chooses to miss on purpose in every round until he is facing a single opponent. The Bad and the Ugly have no reason to shoot the Good, since for the Bad it is suicide and for the Ugly it gives him a 50/50 chance of winning, when he can get $\frac{2}{3}$ by shooting the Bad. Everything happens therefore as if it were a duel only between the Bad and the Ugly, whose winner faces off against the Good. To win this duel, the Bad will shoot at the Ugly. Either he hits and the Good faces him with a probability $q$ of winning, or he misses and faces the Ugly. In this case, the Good has no second shot and only has a 1-in-3 chance of winning. Finally, the probability that the Good wins with this strategy is $p' = \frac{1}{2}q + \frac{1}{2} \times \frac{1}{3} = \frac{5}{12}$. Since $p' > p$, this is the best strategy.

**35** Let us show by induction over $n \geq 1$ that if there are $n$ gas stations, with just enough fuel to go the whole way around, there is a gas station from which a car can leave to complete the lap.

With only one gas station that contains just enough fuel, a car leaving from this gas station can go the whole way around.

Let us suppose the result is true for $n$ gas stations. Let us show it also holds for $n+1$ gas stations. Notice that there is a gas station that has enough gas to make it to the next one. Indeed, if no station had enough gas for the car to reach the next station, all the stations combined would not have enough gas for the car to make the loop. A gas station containing enough gas for the car to reach the next station can be replaced, conceptually, by a gas station containing the combined amount of gas at both these stations. By induction, there exists a gas station from which a car can leave to complete the lap.

There is a direct, but less intuitive proof of this result. Consider a car with a big tank and enough gas to go all the way around by filling up at every gas station. After this lap, it has as much gas as at the start. There was a gas station where its tank was at its least full, and one can notice that this station can be used as the starting point for an empty car.

**36** Let us begin by noting that whether or not the student sees the result of his prediction before guessing the next card doesn't matter. Indeed, since he is aiming for an impeccable guessing sequence, he can always assume the color of the card is the same as he predicted, because otherwise he has already lost and future predictions are irrelevant.

Overall, he must guess the exact distribution of red cards in the pack. We can ask ourselves how many possible distributions there are. The answer is 52 choose 26 (see reminder 2), which is $\frac{52!}{26! \times 26!}$. All these distributions are equally likely. For example, the distribution in which there are only red cards followed by only black cards will happen in $26! \times 26!$ cases since there are 26! ways to arrange the 26 red cards and 26! ways to arrange the black cards. Likewise, any distribution (for example, one red card, one black card, one red card, etc.) will happen in $26! \times 26!$ cases. Thus, since all distributions are equally likely, there is one chance in $\frac{52!}{26! \times 26!}$ of succeeding in the game with, for example, the strategy of guessing red for the first 26 cards and black for the following 26 without even looking at the cards his friend is showing him.

**37** We start by weighing four against four: weights 1, 2, 3, and 4 on the left and weights 5, 6, 7, and 8 on the right. If the weight is the same, we then know that all these weights have a standard mass and the different weight is one of weights 9, 10, 11, or 12. We then weigh three against three: weights 1, 2, 3 on the left and weights 9, 10, 11 on the right. If the weights are the same again, we conclude that weights 9, 10, and 11 are standard and thus that weight 12 is different. All that remains is to weigh it against a standard weight to determine if it is heavier or lighter. If the weight was not the same and if the scale leans left, we deduce that the differing weight is among weights 9, 10, and 11 and that it is lighter. We conclude easily with a final weighing. (The case is symmetric if the scale leans right.)

Let us return to our first weighing. If now the weights aren't the same, and the scale tilts to the left, we can conclude that there are two possibilities for the differing weight:

- The differing weight is among weights 1, 2, 3, and 4 and it is heavier than the standard weights.
- The differing weight is among weights 5, 6, 7, and 8 and it is lighter than the standard weights.

(The case is symmetric if the scale leans right.) We can also conclude that weights 9, 10, 11, and 12 have standard mass.

We weigh 1, 2, and 7 on the left against 3, 4, and 9 on the right. If the weight is the same, we conclude that all these weights have standard mass and the differing weight is either 5, 6, or 8. We then conclude by weighing 5 against 6. If the scales tip left, then the differing weight is either 1 or 2 and we conclude easily with one last weighing. Finally, if the scales tip to the right, the differing weight is 3, 4, or 7, and we conclude by weighing 3 against 4.

Note that the search for such a solution is made easier by using the method for comparing outcomes to hypotheses (see reminder 1). For example, this method allows for the direct elimination of a first three against three weighing: if the weight is the same, there would be 12 solutions (6 weights) left and only 9 possible results (two weighings).

**38** We begin by separating the coins into a group of 25 coins and a group of 75 coins. In the group of 75 coins, there is some number $n$ of heads. We don't know $n$, but we only know that $n$ is between 0 and 25. Therefore, in the group of 25 coins, there are therefore $25 - n$ heads since the total number of heads is 25. Now we flip all the coins in the group of 25 coins. Thus, we change the heads into tails and the tails into heads. At the end of the procedure, there are, in the group of 25 coins, $n$ heads and $25 - n$ tails. The problem is solved.

**39** Whether there are two or seven colors, it is possible to free 99 students. This is optimal because the student at the top of the staircase cannot be freed for certain, since she has no information that would allow her to find the color of her hat.

The idea is to make the answer of the first student useful to the others. She can count the black hats she sees and say "white" if the count is even and "black" if not. From there, all the students will find the color of their hat. Indeed, suppose the first student said "white" (the case is symmetrical if she said "black"). The next student (the one on the $99^{\text{th}}$ step) counts the black hats he sees; if the count is an even number his hat is white, if not it must mean his hat is black. The next student, considering the response of the previous ones, knows the parity of the number of black hats of students on steps 1 to 98, so he can reply too, and so on until the student at the bottom of the stairs has a chance to respond.

And with seven colors? One can number the seven colors from 0 to 6. The first student questioned computes the remainder mod 7 (see reminder

6) and responds by the color corresponding to this number. For example, if the sum is 50, she responds 1 since $50 \equiv 1\,[7]$. In turn, the second student computes the sum of the hats he sees and deduces from the difference the number on his hat. For example, if he computes 48 and the first student announced 1, he says his color is 2 since this is the only number between 0 and 6 which added to 48 gives a remainder of 1 mod 7. The third student then knows the number of the second and proceeds in the same manner. All students deduce in turn the color of their hat.

We could ask ourselves what would happen with infinitely many colors. For this to be meaningful, we can refer to the colors as the integers: students have hats bearing arbitrary integers. It is easy since the first student can give much more information.

For example, she can give the sum of all hats she sees, and students in turn deduce the numbers they're wearing. The first student can even allow each of the others to find the number on their hat at the same time. It is sufficient to find an injection from $\mathbb{N}^{99}$ into $\mathbb{N}$. One way is by using the unicity of prime factorization: let $c_1, c_2, \ldots, c_{99}$ be the numbers worn by the 99 students on steps 1 to 99 and $(p_n)_{n\in\mathbb{N}^*}$ be the sequence of prime numbers; then the top student can announce $p_1^{c_1} \times \cdots \times p_{99}^{c_{99}}$.

**40** We separate the coins into two groups, naughts and crosses, as follows:

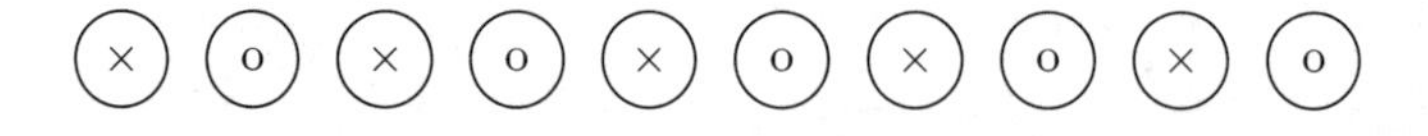

We can choose to collect all naughts or all crosses, it only remains to choose the symbol which is the most profitable.

In the given instance, the sum of crosses is worth $2 + 7 + 6 + 5 + 9 = 29$ and the sum of naughts is worth $4 + 1 + 10 + 8 + 3 = 26$. One must thus choose the crosses and begin by taking coin 2. Then if the opponent chooses coin 4 we take coin 7 and if he takes coin 3 we take 9, and so on.

Note also that the sum of the coins is worth $1 + 2 + 3 + 4 + 5 + 6 + 7 + 8 + 9 + 10 = 55$, an odd number, and thus the sum on the crosses cannot be equal to the sum on the naughts. Thus, we can be certain of collecting strictly more than the gamemaster.

**41** In fact, one can find two opposite points where the temperature is the same on any meridian. Let us place ourselves on an arbitrary meridian. Consider the function that to a point on this meridian maps the difference of temperature between this point and its opposite point. One must show that this function, which is continuous, has a zero. When we travel along the meridian from the North Pole to the South Pole, the function changes sign (its value at the North Pole is the opposite of its value at the South Pole). Thus, by the intermediate value theorem, there is a point where the difference is equal to zero.

Note that there exists a much stronger result: there are always two exact opposite points on Earth where the temperature and air pressure are the same. This is the Borsuk-Ulam theorem in two dimensions. This theorem, proven in 1933 by Karol Borsuk, ensures that for any continuous function from an $n$-dimensional sphere (that is to say a sphere in $\mathbb{R}^{n+1}$) into $\mathbb{R}^n$, there are two antipodal points which have the same image by $f$. We proved above the theorem in one dimension.

**42** This seems impossible. If all the rooms are busy, it is hard to see how we could find more room by rearranging customers. Of course, in a finite hotel this is impossible. But an infinite hotel possesses strange properties. Let us see how to free a room.

It suffices to ask every customer to move along by one room. The customer in room 1 goes to room 2, the customer in room 2 goes to room 3 and so on: the customer of room $n$ will move into room $n + 1$. At the end of this move, room 1 is empty: we freed a room!

Once we freed a room, we would think that it would suffice to repeat this operation infinitely many times to free up infinitely many rooms, except it is impossible! Let us reason by contradiction: if we repeat this operation infinitely many times then all the rooms would be empty and there would be no customers left. It is absurd; we must find another method.

It is sufficient to ask the customer in room 1 to go to room 2, the customer in room 2 to go to room 4, and the customer in room $n$ to go to room $2n$. After this move, all odd numbered rooms will be empty and it only remains to move the new customers into them. The new customer 1 will go in room

1, the new customer 2 will go in room 3, and so on: the new customer $n$ will go to room $2n-1$.

**43** We observe that the last free seat can only be the first passenger's or the last passenger's, otherwise it would have been taken. The passengers, when sitting, do not differentiate between these two possibilities: we could swap the numbers on the seats of the first passenger and the last passenger, and it would not change the boarding of the 98 other passengers. The hundredth passenger thus has a 50/50 chance of finding their seat empty.

**44** Imagine, to begin with, that there are only two pirates and they have 100 doubloons to share. The oldest makes the first proposition: 100 doubloons for him and nothing for the other. This distribution is necessarily accepted since even though the second pirate can oppose it, there cannot be a majority against it.

Let us now move to the case where there are three pirates and 100 doubloons. Whatever the distribution the eldest proposes, the second pirate will oppose it. Indeed, if the distribution is refused, there will be only two pirates left and he will be the eldest. He will propose everything for himself and nothing for the other and will win. On the other hand, the third pirate gains to accept the distribution of the eldest since he knows he will get nothing if it doesn't pass. Thus, it suffices for the oldest pirate to offer 99 doubloons for himself and 1 doubloon for the third pirate. The third will accept: 1 doubloon is always better than none. The second will oppose it, but the distribution will pass anyway.

With four pirates, the first proposes 99 doubloons for himself, 1 doubloon for the third, and nothing for the other two. The third is in favor since he knows that if there is a mutiny, he will end up in second position among three pirates and will get nothing. The distribution is thus accepted.

We obtain in this manner the distributions for 4, 5, ..., 10 pirates:

| Order of Age / No. of pirates | 1 | 2 | 3 | 4 | 5 | 6 | 7 | 8 | 9 | 10 |
|---|---|---|---|---|---|---|---|---|---|---|
| 2 | | | | | | | | | 100 | 0 |
| 3 | | | | | | | | 99 | 0 | 1 |
| 4 | | | | | | | 99 | 0 | 1 | 0 |
| 5 | | | | | | 98 | 0 | 1 | 0 | 1 |
| 6 | | | | | 98 | 0 | 1 | 0 | 1 | 0 |
| 7 | | | | 97 | 0 | 1 | 0 | 1 | 0 | 1 |
| 8 | | | 97 | 0 | 1 | 0 | 1 | 0 | 1 | 0 |
| 9 | | 96 | 0 | 1 | 0 | 1 | 0 | 1 | 0 | 1 |
| 10 | 96 | 0 | 1 | 0 | 1 | 0 | 1 | 0 | 1 | 0 |

With 10 pirates, the first proposes 96 doubloons for himself, nothing for all the even position pirates, and 1 doubloon each for the others.

**45** The first question to ask oneself is: is it even possible for the baboon to run at an irregular pace? The answer is yes! For example, it can run 500 meters at a regular pace of 5 m/s, then the remaining 500 meters at a regular pace of 2 m/s, then again 500 meters at 5 m/s, then another 500 meters at 2 m/s and so forth.

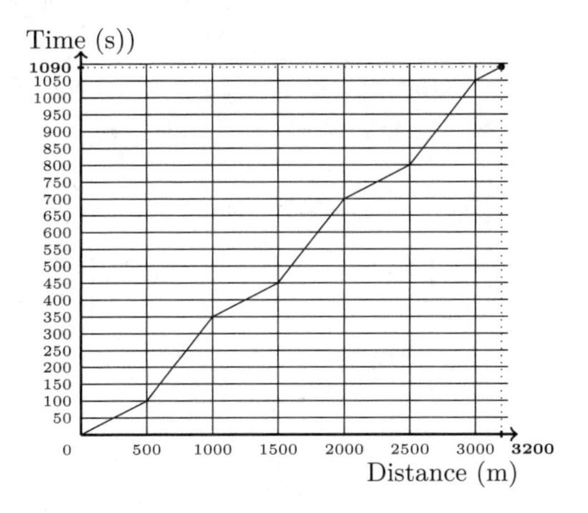


By running in this manner, if we consider an arbitrary 1 km interval, we see that the baboon will have run 500 meters at 5 m/s, which will take 100 seconds and 500 meters at 2 m/s, which will take 250 seconds. Hence, it will have covered this kilometer in 350 seconds, which is 5 minutes 50 seconds. At this rate, the baboon finishes the race in 18 minutes 10 seconds, whereas the chimpanzee finishes the race in 18 minutes 24 seconds. The baboon can thus win the race. We note that in the limit case, in which the baboon can run infinitely fast, the baboon can finish the race in 17 minutes 30 seconds.

**46** Snow White will weigh simultaneously one ingot of the first dwarf, 2 ingots of the second dwarf, 4 ingots of the third dwarf, 8 ingots of the fourth dwarf, 16 ingots of the fifth dwarf, 32 ingots of the sixth dwarf, and 64 ingots of the seventh dwarf. The total should weigh 12,700 grams. We compute the difference between 12,700 grams and the result of the weighing. We express this difference in grams, write this number in base 2, and read off who are the thieves.

For example, if the result of the weighing is 12,670 grams, the difference is 30 grams, which is $0011110_2$ in base 2 (see reminder 7). Thus, we see that the thieves are dwarves 2, 3, 4, and 5.

**47** We could try to make a triangle that shares a whole side with the square, and two other triangles that share half of opposite edges of the square. On the figure below, these triangles are $(AJD)$, $(AIJ)$, and $(DJK)$. $J$ must be placed outside the half-disks contained in the dotted semi-circles whose diameters are $[AD]$, $[AI]$, and $[DK]$. Mirroring the figure on the other side, creating $J'$ and its three mirrored triangles and connecting $J'$ and $J$ to $I$ and $K$, it remains to check that the triangles are all acute.

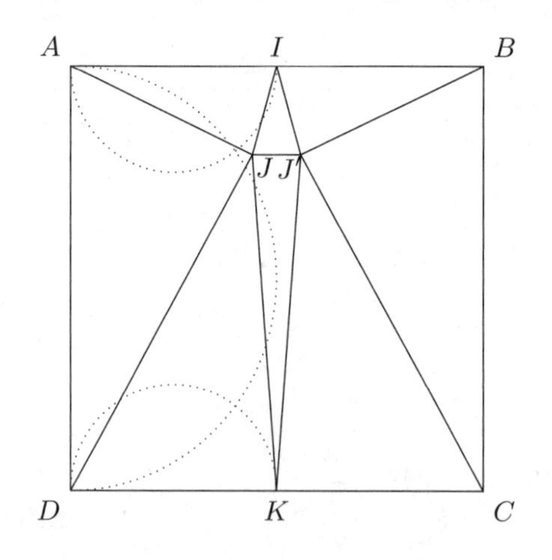

**48** It is impossible. Let us assume for a contradiction that we managed to tile the bathroom. We paint the tiles as follows:

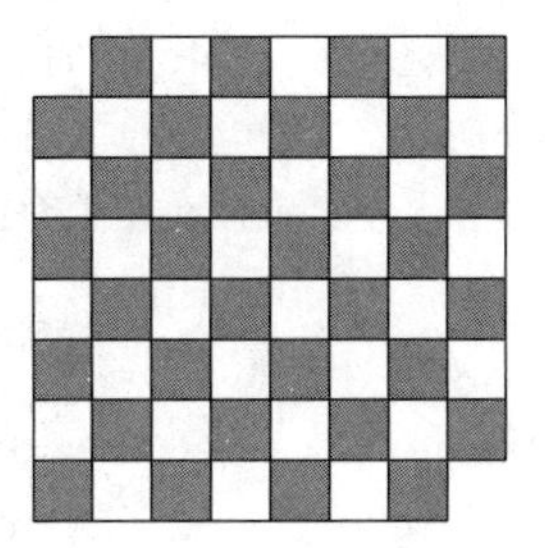

We notice that there are 30 white squares and 32 black squares. But each domino covering the floor has had one tile painted black and the other painted white. There should therefore be the same number of white tiles as black tiles. A contradiction.

**49** It may seem surprising, but the older brother has, despite his extra coin, only a 50/50 chance of getting strictly more tails than his younger brother.

There are only two possible cases: either the older brother has strictly more tails than his brother, or strictly more heads. At least one of these two cases will occur: since the older brother has one more coin, he cannot have both a number of heads less than or equal to his younger brother and a number of tails less than or equal to his younger brother. Moreover, both these cases cannot occur at once: since he only has one extra coin, the older brother cannot have both strictly more heads and strictly more tails than his younger brother. We will always have one case or the other, but never both at the same time.

Both cases are symmetric; thus, they have the same probability, and since they are complementary, their probability is $\frac{1}{2}$. Thus, the older brother only has a 50/50 chance of getting strictly more tails than his younger brother.

**50** The younger brother is right: the game is not fair. It can seem surprising after the older brother's convincing reasoning. Nevertheless, there are two mistakes in this line of reasoning.

**First mistake** It is true that, on three throws, the probability of obtaining THH is equal to the probability of obtaining HHT and is equal to $\frac{1}{8}$. Nevertheless, this does not imply that the mean occurrence time of THH and HHT is 8. It is indeed the case, but purely by chance. In fact, we can show that the mean occurrence time of TTT is 14 and not 8. The computations for the mean occurrence times of TTT and THH are a bit heavy. Let us rather show that the mean occurrence times of TT and TH are different even though the probability of appearing on two throws is the same.

Let us note $t_0$ as the mean occurrence time of TT and $t_1$ as the mean occurrence time of TT once we've gotten T. Once we have obtained a T, one of two possibilities can occur:
– We get heads and we are back in a situation in which the mean occurrence time is $t_0$, in which case the mean time since the first H is $t_0 + 1$.
– We get a second T and it is over; the mean time since the first T is then 1. Thus

$$t_1 = \frac{1}{2}(t_0 + 1) + \frac{1}{2} \times 1.$$

In the same manner, when we start from 0 (H or no previous throw), we have a 50/50 chance of obtaining T, and a 50/50 chance of H; hence

$$t_0 = \frac{1}{2}(t_0 + 1) + \frac{1}{2}(t_1 + 1)$$

It follows that $t_0 = 6$.

Let us compute in the same manner the mean occurrence time of TH. We note $t'_0$ as the mean occurrence time, and $t'_1$ as the mean occurrence time starting at T. This time, by starting at T, if we fail H, we restart at T rather than 0.

$$t'_1 = \frac{1}{2}(t'_1 + 1) + \frac{1}{2} \times 1$$
$$t'_0 = \frac{1}{2}(t'_0 + 1) + \frac{1}{2}(t'_1 + 1)$$

By replacing, $t'_0 = 4$, we indeed have $t'_0 \neq t_0$.

**Second mistake** It is true that the mean occurrence times are the same, but it doesn't imply the game is fair. Let us assume that THH appears 9 times out of 10 on the first throw, and 1 time out of 10 on the 21st throw;

let us also assume that the sequence HHT always occurs on the third throw. The mean occurrence time of THH is $\frac{9}{10} \times 1 + \frac{1}{10} \times 21 = 3$; the one of HHT is also 3, but 9 out of 10 times, THH appears before HHT. Of course, it is utterly unrealistic that HHT always appears on the third throw, but this example shows that basing an argument only on the mean occurrence time is not sufficient to draw a conclusion. An event with a very large mean occurrence time can very well precede 9 times out of 10 another event that has a short mean occurrence time.

Let us now see in more detail why the older brother has an advantage. It is because HHT starts with the end of THH. When HHT appears, unless there has only been H before, THH will have appeared earlier. Indeed, it is sufficient that there be an H at the start for the older brother to ensure victory. There are four possibilities for the first two throws: TT, TH, HT, HH. If the first two throws are TT, TH, or HT, the older brother is sure to win. If HHT appears, it will have been preceded by THH. If not, if the first two throws are HH, the younger brother is sure to win: he just waits for the coin to be T. In the end, the younger brother only has a 1-in-4 chance of winning, against his older brother's 3-in-4 chance.

**51** The probability that the younger brother wins against his older brother is given by the following table:

| younger \ older | TTT | TTH | THT | HTT | HHH | HHT | HTH | THH |
|---|---|---|---|---|---|---|---|---|
| TTT | | $\frac{1}{2}$ | $\frac{2}{5}$ | $\frac{1}{8}$ | $\frac{1}{2}$ | $\frac{3}{10}$ | $\frac{5}{12}$ | $\frac{2}{5}$ |
| TTH | $\frac{1}{2}$ | | $\frac{2}{3}$ | $\frac{1}{4}$ | $\frac{7}{10}$ | $\frac{1}{2}$ | $\frac{5}{8}$ | $\frac{2}{3}$ |
| THT | $\frac{3}{5}$ | $\frac{1}{3}$ | | $\frac{1}{2}$ | $\frac{7}{12}$ | $\frac{3}{8}$ | $\frac{1}{2}$ | $\frac{1}{2}$ |
| HTT | $\frac{7}{8}$ | $\frac{3}{4}$ | $\frac{1}{2}$ | | $\frac{3}{5}$ | $\frac{1}{3}$ | $\frac{1}{2}$ | $\frac{1}{2}$ |
| HHH | $\frac{1}{2}$ | $\frac{3}{10}$ | $\frac{5}{12}$ | $\frac{2}{5}$ | | $\frac{1}{2}$ | $\frac{2}{5}$ | $\frac{1}{8}$ |
| HHT | $\frac{7}{10}$ | $\frac{1}{2}$ | $\frac{5}{8}$ | $\frac{2}{3}$ | $\frac{1}{2}$ | | $\frac{2}{3}$ | $\frac{1}{4}$ |
| HTH | $\frac{7}{12}$ | $\frac{3}{8}$ | $\frac{1}{2}$ | $\frac{1}{2}$ | $\frac{3}{5}$ | $\frac{1}{3}$ | | $\frac{1}{2}$ |
| THH | $\frac{3}{5}$ | $\frac{1}{3}$ | $\frac{1}{2}$ | $\frac{1}{2}$ | $\frac{7}{8}$ | $\frac{3}{4}$ | $\frac{1}{2}$ | |

We notice with this table that, whichever sequence was chosen by the older brother, the younger brother can in fact choose a sequence which makes him the winner on average. For example, if the older brother chose THT, the younger brother could choose TTH or HHT: in both cases, the younger brother will have more than a 50/50 chance of winning. Thankfully, he doesn't need to compute everything. The older brother can propose only eight sequences: TTT, TTH, THT, HTT, HHH, HHT, HTH, THH. Up to changing T into H and H into T, we can restrict ourselves to the study of the first four sequences: TTT, TTH, THT, THH. Let us take these cases one by one.

**TTT** The idea is to choose a sequence whose end is the start of the one proposed by the older brother. A sequence that might seem clever is HTT. Indeed, we see that if the first three throws are not TTT, the younger brother wins because if TTT appears, HTT will have appeared first. He thus has a 7-in-8 chance of winning.

**TTH** If the older brother proposes TTH, we saw in the previous riddle that HTT allows the younger brother to have a probability of winning of $\frac{3}{4}$.

**THT** Here, the younger brother wants a sequence whose end is the start of his older brother's, so he can choose TTH or HTH. However, it is clear that with HTH, he only has a 50/50 chance of winning since T and H are symmetric. Let us then compute the probability $p$ of winning with TTH against THT. To help ourselves in this, we will make use of a tree of tosses, and we can assume that the first toss is T since any previous H doesn't change anything.

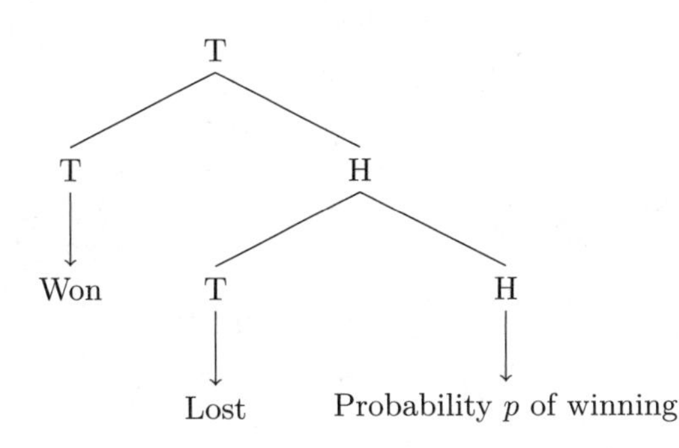


We thus have $p = \frac{1}{2} + \frac{1}{4}p$; hence, $p = \frac{2}{3}$ as the table showed.

**HTT** A sequence granting a greater than 50/50 chance of winning against HTT is HHT. The probability $p$ of winning with HHT against HTT is computed in the same way that the TTH against THT case was computed. The tree is almost the same.

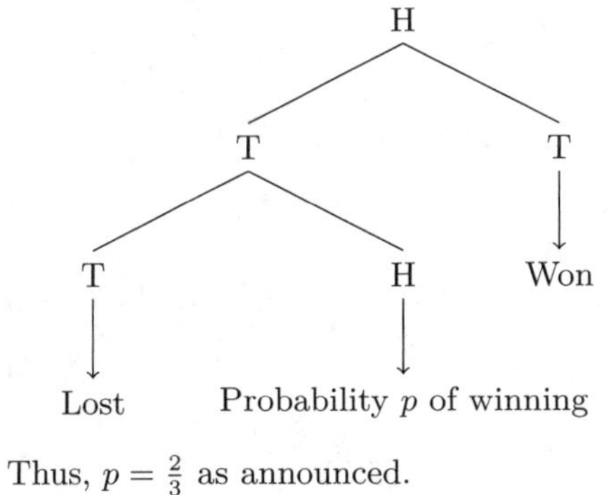


Thus, $p = \frac{2}{3}$ as announced.

**52** The key is to spin the problem differently: with at most $n$ drops, how many floors can one test? We notice already that as soon as the first plate breaks, one must test all the forgotten floors one by one with the remaining plate, starting by the lowest and going up: there is no other approach when one has only one plate left.

Assume we can't do more than $n$ throws. We can drop the plate from the $n$th floor at most. Indeed, if it breaks, one will have to test in the worst case floors 1, 2, 3, ..., $n-1$ with the second plate. If it doesn't break, then there are two plates left and $n-1$ trials, so we can drop the plate from the $n+(n-1) = (2n-1)$th floor and if it breaks there are $n-2$ trials to test $n-2$ floors. Thus, with two plates and $n$ trials, we can test at most $u_n = n + (n-1) + (n-2) + \cdots + 1$ floors.

$$u_n = \sum_{k=1}^{n} k = \frac{n(n+1)}{2}$$

For 100 stories, we have $u_{13} < 100 < u_{14}$. It will thus take 14 drops in the worst case.

## 53

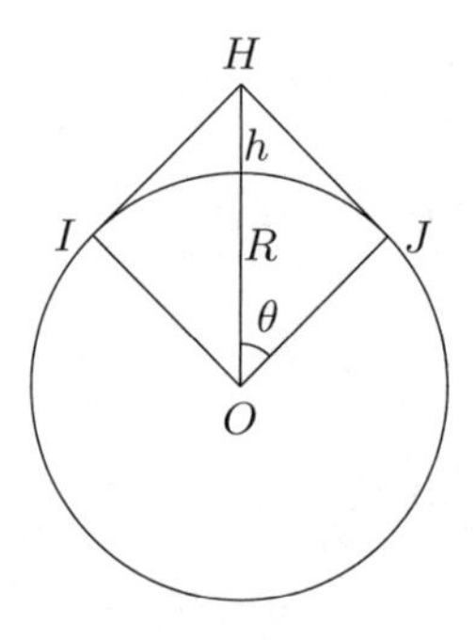


The new length of the cable is $IH + HJ + \overset{\frown}{IJ}$, where $\overset{\frown}{IJ}$ is the length of the arc from $I$ to $J$ via the other side of the circle. Given the angle of $2\theta$, the length of this arc from $I$ to $J$ is $\overset{\frown}{IJ} = 2\pi R - 2\theta R$. Moreover, $IH = HJ = R\tan(\theta)$ where $\theta$ is given by $\cos(\theta) = \frac{R}{R+h}$.

The cable is now longer than before by a length $l = 1$ meter, so $2IH + \overset{\frown}{IJ} - 2\pi R = l$, and we obtain by substitution: $\tan(\theta) - \theta = \frac{l}{2R}$, which already implies that $\theta$ is very small. We can thus make the following approximations:

$\tan(\theta) - \theta \simeq \frac{\theta^3}{3}$, $\cos(\theta) \simeq 1 - \frac{\theta^2}{2}$, and $\frac{R}{R+h} = \frac{1}{1+\frac{h}{R}} \simeq 1 - \frac{h}{R}$, thus $\frac{h}{R} \simeq \frac{\theta^2}{2}$. In the end, we obtain $\frac{h}{R} \simeq \frac{1}{2}\left(\frac{3l}{2R}\right)^{\frac{2}{3}}$, which is

$$h \simeq \frac{R^{\frac{1}{3}}}{2}\left(\frac{3l}{2}\right)^{\frac{2}{3}}$$

Numerically, $h \simeq 122\,\text{m}$. This surprising result was well worth doing a bit of geometry! Without doing any approximations, we can use formal computation software to find $h \simeq 122\,\text{m}$, since the approximations made previously were excellent.

**54** Let us denote $b_n$ as the average number of loops that the young man creates with a plate of $n$ spaghetti. Thus, $b_1 = 1$, and we seek $b_{100}$. Let us consider a plate of spaghetti with $n \geq 2$. For his first knot, the young man either ties a spaghetti to itself or the two spaghetti to each other. In the first case, the spaghetti is eliminated: there will be on average $b_{n-1} + 1$ loops. In the second case, everything happens as if the plate only had $n - 1$ spaghetti on it: there will be on average $b_{n-1}$ loops. The probability that the first spaghetti tied formed a loop is $\frac{1}{2n-1}$: indeed the young man chose the first end at random and then has $2n - 1$ choices for the second. The average number of loops is thus

$$\begin{aligned} b_n &= \frac{1}{2n-1}(b_{n-1}+1) + \left(1 - \frac{1}{2n-1}\right) b_{n-1} \\ &= b_{n-1} + \frac{1}{2n-1} \end{aligned}$$

By an obvious induction, $b_n = \sum_{k=1}^{n} \frac{1}{2k-1}$. Numerically, $b_{100} \simeq 4.4$.

One can show that $\frac{2b_n}{\ln(n)} \underset{n\to+\infty}{\longrightarrow} 1$ which means that for large $n$, $\frac{\ln(n)}{2}$ is a good approximation of $b_n$. We say that $b_n$ is on the order of $\frac{\ln(n)}{2}$. The growth of $b_n$ is logarithmic, and thus much slower than one might have thought.

**55** We notice that for a locker to remain closed after all the passes, the child must have changed its state an even number of times. Thus, she must have passed this locker an even number of times. In other words, the locker's number must have an even number of factors, not counting 1, so an odd number of factors in total. This is the case if and only if it is a square. Indeed, let $n$ be a nonzero integer, and let $f_1, f_2, \ldots, f_k$ be its factors. We can group them by pairs in the following way: we pair factor $f$ with factor $\frac{n}{f}$. If $n$ is not a square, $\frac{n}{f}$ is always different from $f$, so $n$ has an even number of factors. If $n$ is a square, there exists a unique factor $f$ such that $\frac{n}{f} = f$, so $n$ has an odd number of factors. Finally, the lockers that are closed after all passes are those whose numbers are squares: numbers 1, 4, 9, and so forth.

Another method for computing the number of factors of $n$ is to decompose it into prime factors:

$$n = \prod_{i=1}^{k} p_i^{\alpha_i}$$

The number of factors of $n$ is then given by the following formula:

$$f(n) = \prod_{i=1}^{k} (\alpha_i + 1)$$

Thus, $n$ has an odd number of factors if and only if all the $\alpha_i$ are even. But, to say that all the $\alpha_i$ are even is equivalent to saying that $n$ is a square.

**56** We begin by drawing the graph symbolizing the connections between the six people: each person is represented by a vertex numbered 1 through 6, and these connections (knowing or not knowing each other) with the other people are represented by lines (see figure).

Person 1 has five relations with the other people in the group, of which at least three relations are of the same nature, such as knowing other people. On the figure, three of these relations (in bold) are those with persons 3, 4, and 6. There are then two cases:

– If two people among the three people 3, 4, and 6 know each other, then we have a trio of people who know each other pairwise (these two people plus person 1).

– Otherwise, these three people do not know each other pairwise.

In either case, we have in fact a trio who know each other pairwise or do not know each other pairwise.

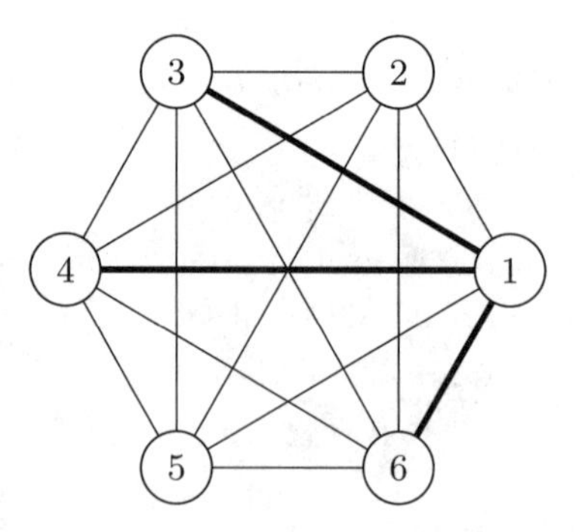

**57** If the first rat tastes a mix of the first 500 bottles, its death or survival the next morning allows the problem to be cut in half. To divide the problem again, the second rat must also taste half the bottles. But we must ensure all rats are useful: if the first rat tastes a mix of the first 500 bottles and the second rat tastes a mix of the last 500 bottles, the second rat is no use. To divide by two the number of bottles as many times as wanted without difficulty, the simplest is to renumber the bottles in base 2 (see reminder 7).

We have $2^9 \leq 999 < 2^{10}$, so one will need 10 digits to number all the bottles. The numbers will go from $0000000000_2 = 0$ (the number of the first bottle) to $1111100111_2 = 999$ (the number of the last bottle). The first rat will taste all the bottles whose last digit is a 1. It will thus taste bottles $0000000001_2$, $0000000011_2$, $0000000101_2$, and so on. The second rat tastes all bottles whose second-to-last digit is a 1 and so forth through the 10th rat, who tastes all the bottles whose first digit is 1. The next day, the butler looks at which rats died. If the $i$th rat is dead, since he tasted all the bottles whose $i$th digit is a 1 (and only those), the butler knows that the number on the poisoned bottle has a 1 in the $i$th position. The butler can in this manner find the number of the poisoned bottle. This method requires 10 rats. Let us show that one can do no better.

We will show that with $n$ rats we cannot discriminate between more than $2^n$ bottles. In other words, if we assume that the butler can determine which bottle is poisoned among $k$ bottles with $n$ rats, then we have $k \leq 2^n$. To do so, we will compare the number of hypotheses with the number of outcomes; this method is detailed in the reminder on the pigeonhole principle (reminder 1). There are $k$ bottles. Either the first bottle is poisoned, or it is the second, ..., or it is the last. There are thus $k$ different possible hypotheses about the solution of the problem. When the butler has the rats taste mixes from different bottles, the next day each rat will be alive or dead. Since there are $n$ rats, there are $2^n$ possible outcomes. By the pigeonhole principle, to identify the bottle we must have more possible outcomes than different solutions to the problem. We thus have $k \leq 2^n$. With $k = 1{,}000$ bottles, we obtain $1{,}000 \leq 2^n$; hence, $10 \leq n$. At least 10 rats are required.

**58** For $k$ between 0 and 100, we note $c_k$ as the average number of cards the girl must buy to finish the collection given she already has $k$ distinct cards. For example, $c_{100} = 0$ and $c_{99} = 100$ since when she has 99 distinct cards, there is only one missing, and since she has a 1-in-100 chance of buying the missing one each time she buys a card, she must still buy 100 cards on average to finish the collection. We seek $c_0$.

Let $k$ be between 0 and 100, and suppose she has $k$ distinct cards. Let her buy one more. This card has a $k$-in-100 chance of being a card she already owns (and in this case she needs to buy $c_k$ more cards) and a $(100-k)$-in-100 chance of being a new card (in which case she only needs to buy $c_{k+1}$ cards). We obtain the following relation:

$$c_k = 1 + \frac{k}{100} c_k + \frac{100-k}{100} c_{k+1}.$$

This relation can be rewritten:

$$c_k - c_{k+1} = \frac{100}{100-k}$$

We sum from 0 to 99:

$$\sum_{k=0}^{99} (c_k - c_{k+1}) = \sum_{k=0}^{99} \frac{100}{100-k}$$

$$\sum_{k=0}^{99} (c_k - c_{k+1}) = c_0 - c_{100} = c_0$$

$$\sum_{k=0}^{99} \frac{100}{100-k} = \sum_{k=1}^{100} \frac{100}{k}$$

We obtain the following:

$$c_0 = \sum_{k=1}^{100} \frac{100}{k}$$

Hence, it will take $c_0 \simeq 518.7$ days on average for the girl to complete her collection.

**59** To find the price of this game is to find how much money it is worth on average, that is, how much one wins on average, which is the expectation denoted $\mathbb{E}(G)$. To compute this expectation, we have to determine all the sums $G_n$ that one could win, as well as the probability $\mathbb{P}(G_n)$ of winning each of them. The expectation is then computed as the sum of all the possible gains multiplied by their associated probability:

$$\mathbb{E}(G) = \sum_{n=0}^{N} \mathbb{P}(G_n) \times G_n$$

In this game, one wins $2^{n-1}$ dollars if the banker lands heads after $n$ tosses. The possible gains are thus $G_n = 2^{n-1}$ for $n \geq 1$.

For $n \geq 1$, one wins $G_n = 2^{n-1}$ dollars if the banker tosses a series of $(n-1)$ tails followed by a heads, which is to say he lands the sequence

$$\underbrace{\text{T...T}}_{(n-1)\text{ times}}\ \text{H}$$

The probability of this event is thus

$$\mathbb{P}(G_n) = \underbrace{\frac{1}{2} \times \ldots \times \frac{1}{2}}_{(n-1)\text{ times}} \times \frac{1}{2} = \left(\frac{1}{2}\right)^n$$

There are infinitely many possible $n$; the sum is thus infinite. Finally, we obtain for the expectation

$$\mathbb{E}(G) = \sum_{n=1}^{+\infty} \mathbb{P}(G_n) \times 2^{n-1} = \frac{1}{2} + \ldots + \frac{1}{2} + \ldots = +\infty$$

We obtain an infinite expectation. This result is paradoxical. Even if one is very lucky, the banker will eventually land on heads at some point and the game will end. In every case, he will only pay out a finite sum of money. Nevertheless, the expected gain is infinite!

The question of the price of the game thus seems unsolvable. To pay an infinite amount of money to play this game is nonsense, and one would be sure to lose an infinite amount of money each time. Indeed, in every outcome, one receives a finite amount of money as payoff. Moreover, if we pay a finite sum to play this game ($10,000, for example), this game favors us. Indeed, the expectation of the gains is infinite; we know that by playing a large number of times we will necessarily end up winning money.

To lift this paradox, we can assume that the banker will no longer be able to pay out after reaching a certain sum. If, for example, the banker cannot pay us more than $2^{30}$ dollars (which is roughly one billion dollars), the expectation becomes:

$$\mathbb{E}(G) = \sum_{n=1}^{30} \mathbb{P}(G_n) \times 2^{n-1} + \mathbb{P}(G_{31}) \times 2^{30}$$

But $\mathbb{P}(G_{31})$ has now changed:

$$\mathbb{P}(G_{31}) = 1 - \sum_{n=1}^{30} \mathbb{P}(G_n) = 1 - \sum_{n=1}^{30} \left(\frac{1}{2}\right)^n = \left(\frac{1}{2}\right)^{30}$$

One obtains

$$\begin{aligned}\mathbb{E}(G) &= \sum_{n=1}^{30} \mathbb{P}(G_n) \times 2^{n-1} + \mathbb{P}(G_{31}) \times 2^{30} \\ &= \sum_{n=1}^{30} \left(\frac{1}{2}\right)^n \times 2^{n-1} + \left(\frac{1}{2}\right)^{30} \times 2^{30} \\ &= \underbrace{\frac{1}{2} + \dots + \frac{1}{2}}_{30 \text{ times}} + 1 = 16\end{aligned}$$

Thus, the price of the game in these conditions is finite and the paradox is lifted.

A more general approach to lift this paradox is to consider a utility function. The gist of it is to say that when one gets \$2,000 instead of \$1,000, one is twice as happy, but if one receives two billion dollars rather than one billion dollars, one's happiness doesn't change much. We will thus compute the expectation by replacing the gains $G_n$ by the utility of the gains $u(G_n)$. To do so, we need to choose a utility function $u$ that maps a sum of money $x$ to its utility $u(x)$.

A utility function is always increasing and concave. A natural choice for a utility function is the natural logarithm. Let's see what the computation of the expectation looks like with this utility:

$$\begin{aligned}\mathbb{E}(G) &= \sum_{n=1}^{+\infty} \mathbb{P}(G_n) \times \log(2^{n-1}) \\ &= \log(2) \times \left( \sum_{n=1}^{+\infty} \mathbb{P}(G_n) \times (n-1) \right) \\ &= \log(2)\end{aligned}$$

Once again, here we found a finite price and the paradox is lifted. Note that this approach is a generalization of the previous one since the function

$$u : \begin{cases} \mathbb{R}^+ & \to \mathbb{R}^+ \\ x & \mapsto \begin{cases} x & \text{if } x \leq 2^{30} \\ 2^{30} & \text{otherwise} \end{cases} \end{cases}$$

is increasing and concave. Hence, it is a utility function, and it can be used to find the previous result.

**60** Given that there is no possible equality, one of the two players (either the one who starts, or the other) has a winning strategy. Indeed, either the first player has a winning strategy, or the second player can always beat the first player regardless of the first player's first move. In the second case, the second player has a winning strategy.

Assume, for a contradiction, that the second player has a winning strategy. If the first player begins by taking the square in the top right, the second player will find a move that, despite the first player's best efforts, will lead to the second player's victory. But, since the first move consisted only in taking the first square, the second move could have been played from the start by the first player. But then, in this situation, the second player is blocked; regardless of what the second player does, the first player will win (if they play optimally). This contradicts the fact that he has a winning strategy! It is thus the first player who possesses a winning strategy.

**61** There are no winning strategies: if both players play well, none will manage to select three numbers whose sum is 15 and it will always be a draw.

What are the different ways to obtain 15? Let us write them all down:

$$9 + 5 + 1$$
$$9 + 4 + 2$$
$$8 + 6 + 1$$
$$8 + 5 + 2$$
$$8 + 4 + 3$$
$$7 + 6 + 2$$
$$7 + 5 + 3$$
$$6 + 5 + 4$$

Consider now the following magic square:[3]

$$\begin{matrix} 2 & 7 & 6 \\ 9 & 5 & 1 \\ 4 & 3 & 8 \end{matrix}$$

What is magical about it is that the sums along each line, each column, and each diagonal are always 15. This gives eight different ways of making 15. But since there are only eight ways to make 15, it suffices to select three numbers on the same line, column, or diagonal. It plays like Tic-Tac-Toe! It is not complicated to check that for Tic-Tac-Toe, there is no winning strategy, hence proving the result we claimed.

**62** The ant will catch up to the car, but after an enormous amount of time!

The first challenge is to determine the equation of motion. Let $(Oz)$ be the axis along which the car moves; thus, the pole is at $z = 0$ and the car is initially at $z = 1\,\mathrm{m}$, which we note as $d$. At time $t$, the car is at $Z(t) = Vt+d$, with $V = 200\,\mathrm{km/h}$. Let $z(t)$ be the position of the ant on the $(Oz)$ axis at time $t$.

[3] A magic square is a square array of positive integers that sum to the same value along any row, column, or main diagonal.

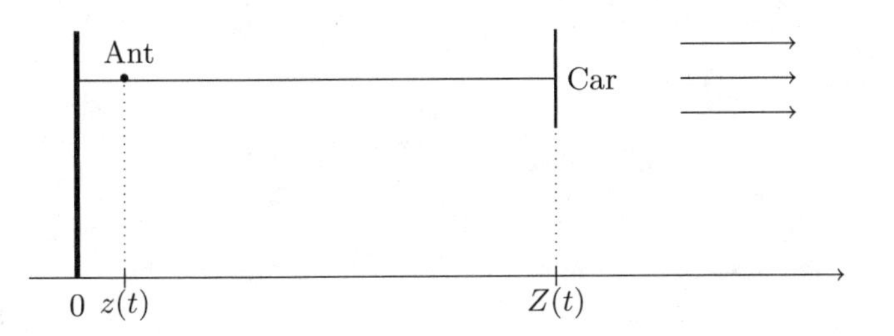


The ant's speed is its relative speed $v = 1\,\mathrm{m/min}$, plus the speed of the point of the elastic upon which it stands at time $t$. What is the speed of an arbitrary point of the elastic? At the pole it is 0, at the car it is $V$. Let us determine the speed of the elastic for a point that is at a certain ratio $x$ of the elastic: $x$ is the ratio of the distance between the point and the pole to the distance between the car and the pole. Since the elastic extends uniformly, a point situated at ratio $x$ stays at ratio $x$; by denoting $\tilde{z}(t)$ as its position at time $t$, $\frac{\tilde{z}(t)}{Z(t)}$ remains constant and equal to $x$. By differentiating $\tilde{z}'(t) = xV$, in particular, for the ant that is at a ratio $x = \frac{z(t)}{Z(t)}$, the speed from the elastic is $\frac{z(t)V}{Z(t)}$. In the end,

$$z'(t) = v + \frac{z(t)V}{Z(t)}$$

$$z'(t) - \left(\frac{V}{Z(t)}\right) z(t) = v$$

This a first order inhomogeneous ordinary differential equation. The solutions of the homogeneous equation (replacing $v$ by 0) are the $Az_0$ with $A$ a constant and

$$z_0(t) = \frac{Vt + d}{d}$$

It suffices now to find a particular solution to have all solutions. Let us seek it with an answer of the form

$$z(t) = \lambda(t) z_0(t)$$

with $\lambda$ a continuously differentiable function. Also, $z$ is a solution if and only if

$$\lambda'(t) z_0(t) = v$$

$$\lambda'(t) = \frac{vd}{Vt + d}$$

Thus, a suitable function is

$$\lambda(t) = \frac{vd}{V} \ln\left(t + \frac{d}{V}\right)$$

In the end, the solutions are

$$z(t) = (Vt + d)\left(\frac{v}{V} \ln\left(t + \frac{d}{V}\right) + C\right)$$

with $C$ a constant. Since $z(0) = 0$, $z(t) = \frac{v}{V}(Vt + d)\ln\left(\frac{V}{d}t + 1\right)$.
The ant catches up to the car if there is $t > 0$ such that $z(t) = Z(t)$. This is equivalent to

$$\frac{v}{V} \ln\left(\frac{V}{d}t + 1\right) = 1$$

or

$$t = \frac{d}{V}\left(\exp\left(\frac{V}{v}\right) - 1\right)$$

The ant thus catches up to the car, after a time $t \simeq 2.5 \times 10^{1438}$ years. Relative to this length of time, the lifetime of the universe is completely insignificant.

**63** We consider a distribution of weights that works, which means it is such that whatever nut the squirrel sets aside, it can always separate the 10 others into two groups of five nuts with the same mass. We want to show that all the nuts have the same mass. From now on, we will assimilate the weights of the nuts to the integers representing these weights in thousandths of a gram.

First, we notice that all the weights have the same parity. Indeed, if a nut has an even weight, since we can separate the other 10 nuts into two groups of the same weight, the sum of the other weights is even. The total sum of nuts is thus even. Likewise if a nut has odd weight. If we now take an arbitrary nut, since the sum of the 10 other weights is even and the total sum is even (respectively odd), this nut has an even (respectively odd) mass.

Now we notice that if a distribution works and contains an even weight, we can divide all the weights by two and it will still work. We can thus divide all the weights of a unit. By removing the minimum mass from every nut, we have a distribution that works with at least one zero mass. Either all the masses are zero and were all equal, or there is at least one nonzero mass. In this case, since there is an even mass (zero), we can divide by two, and so on as many times as necessary to render the nonzero mass odd. We thus have a zero weight and an odd weight: a contradiction. Hence, we have proven our result.

**64** Let $(ABC)$ be the triangle that will model the cake. For any triangle, there is a foot to one of its altitudes. Here we will consider $H$, the foot of the altitude from $B$. Let $I$ and $J$ be the midpoints of line segments $[AB]$ and $[BC]$. Two strokes of the knife following lines $(HI)$ and $(HJ)$ solve the problem. Indeed it is simple to rearrange the pieces cut in this manner for the triangle symmetric to $(ABC)$.

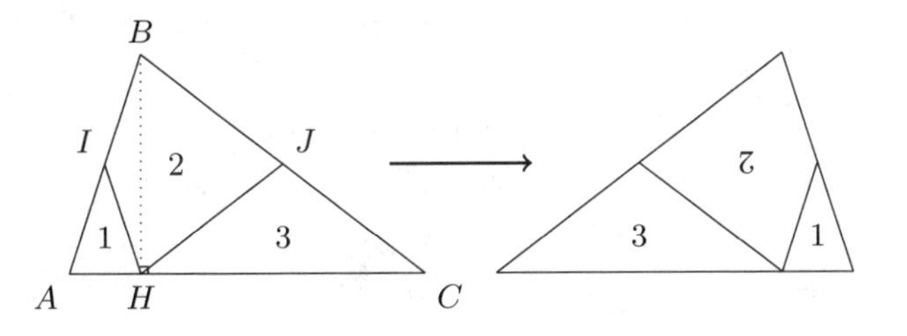


**65** The case $n = 2$ is known to anyone who has at least one sibling. The first sibling cuts the pizza in half and the second chooses their slice. Let us solve the problem of the case $n = 3$. A person cuts what seems to them like one-third of the pizza. If the second person thinks this slice is more than one-third, they can cut some off. The third person may do the same (potentially cutting some off for the second time). The rule is the following: the last person to touch the slice must eat it. Thus, no one is going to cut a slice so that it seems to them like it is less than one-third in fear of ending up with the slice if no one else touches it. Everyone is happy, the last person to touch the slice gets what they think is more than one-third, and the other two don't think that slice is more than one-third, so they think they have at least two-thirds to share. They share the remainder without issues (by the method for the $n = 2$ case).

This reasoning extends easily for $n$ people. The first person cuts the slice they like, then each of the others in turn can cut this slice further, knowing that the last one to cut the slice gets it. We then reduce to sharing between $n-1$ people. We thus have an induction and the problem is solved.

**66** Notice first of all that with two soldiers the result is false: the two soldiers watch each other, and no soldier is left unwatched. Likewise with four soldiers, it is easy to find a configuration where no soldier is left unwatched. Finally, the proposed result is not true for any even number of soldiers, but it is true for any odd number of soldiers.

Let us show the result by induction. For all $n \geq 0$, denote $P_n$ as the property: "With $2n+1$ soldiers, at least one soldier is left unwatched." The initialization with $n = 0$ is trivial. For the induction step, let $n \geq 0$ and let us assume $P_n$ holds. We want to show $P_{n+1}$. We thus consider a group of $2n+3$ soldiers. The mutual distances between soldiers are all different, so there are two soldiers whose mutual distance is minimal. There are two cases. The first case is that one of these two soldiers is watched by another soldier in addition to being watched by the soldier closest to them. In this case, one of the soldiers is unwatched, since if all soldiers were watched then they would all be watched by exactly one soldier. In the other case, the pair of soldiers watch each other but no one else watches either of them. In this case, we can remove them and consider the rest of the group, which is of size $2n+1$, to which $P_n$ applies. There is at least one soldier who is unsupervised and remains unsupervised when we add back in the pair of soldiers we removed since they only watch each other; this gives us the induction.

**67** First, let us determine the minimum number of locks that the locksmith must use. We consider a group of six pirates. This group cannot open the chest alone; thus, there is at least one lock that this group can't open. Now, if we consider any other group of six pirates, there is at least one lock this group can't open, but it differs from the lock of the previous group. Indeed, if two groups of six pirates can't open the same lock, then we can find a group of seven pirates who can't open it either, which is a contradiction. So, we have an injection from the set of groups of six pirates into the set of locks. One must thus use at least $\binom{13}{6} = 1{,}716$ locks.

To put such a system in place, the locksmith numbers the locks from 1 to 1,716 and numbers the possible groups of seven pirates from 1 to 1,716, by exploiting the equality $\binom{13}{7} = \binom{13}{6}$ (see reminder 2). Then, for each of the 1,716 groups, the locksmith hands out to each pirate in group $i$ a key to lock $i$. Let us check if this system works. Any group of six pirates cannot open the chest since they lack the key of the group of seven pirates not present. On the other hand, as soon as seven pirates are present, they own the $i$ key (for any $i$) since the group number $i$ must contain at least one pirate of the group.

**68** First of all, notice that starting with one weight of 1 kg and another weight of 3 kg, we can already determine the weight of all objects between 1 kg and 4 kg. To weigh an object of 2 kg with these weights, it suffices to place the object with the 1 kg weight on the left and the 3 kg weight on the right.

If we choose weights of mass 1 kg, 3 kg, 9 kg, 27 kg, ..., $3^9$ kg, we can determine the weight of any object between 1 kg and $1 + 3 + \cdots + 3^9 = \frac{3^{10}-1}{2}$ kg. Indeed, let us show by induction on $n \geq 0$ that with weights of mass 1 kg, 3 kg, ..., $3^n$ kg, we can determine the weight of any object between 1 kg and $\frac{3^{n+1}-1}{2}$ kg.

Initialization is simple. Suppose the proposition holds for $n \geq 0$. We have weights of mass 1 kg, 3 kg, ..., $3^{n+1}$ kg. If we want to weigh a mass between 1 kg and $\frac{3^{n+1}-1}{2}$ kg, doing so is straightforward; we don't need the $3^{n+1}$ kg weight. To weigh an object whose mass is between $\frac{3^{n+1}+1}{2}$ kg and $\frac{3^{n+2}-1}{2}$ kg, we begin by subtracting $3^{n+1}$ kg from this mass by placing the corresponding weight on the other side. This brings the mass down to between $-\frac{3^{n+1}-1}{2}$ kg and $\frac{3^{n+1}-1}{2}$ kg, which we can weigh with the other weights, which completes the induction.

Can we do better? No. We use the comparison method for outcomes and hypotheses (see reminder 1). Let us compute the maximum number of different weighings we can produce with 10 weights. For each weight we have three possibilities: place it on the left side, place it on the right side, or do not use it at all. This gives $3^{10}$ possible weighings. However, we actually have only $3^{10} - 1$ weighings, since not placing anything on the scales does not count as a valid weighing. We must then divide the result by 2 since the plates are symmetric. We thus obtain at most $\frac{3^{10}-1}{2}$ possible results from weighing. This maximum is reached with the previous system of weights; hence the result is proven.

**69** The elephant can carry at most 533 bananas to the oasis. There are several possible strategies.

**One possible strategy** The elephant carries as many bananas as possible, kilometer by kilometer. It starts from kilometer 0 with 1,000 bananas on its back, and arrives at the first kilometer with 999, drops 998 off, and then returns to the start with the last banana. It does this again and drops off 998 at the first kilometer once more. Finally, it returns to pick up the last 1,000 bananas and drops 999 off at the first kilometer mark. All in all, the elephant ate five bananas to carry $998 \times 2 + 999 = 2{,}995$ bananas the first kilometer. It starts over now and carries 2,990 bananas to the second kilometer mark. For each kilometer, the elephant consumes five bananas to carry all the others. The elephant continues in this way until the 200th kilometer, to which it brings 2,000 bananas. It leaves the 200th kilometer with 1,000 bananas on its back, drops 998 off at the 201st kilometer, keeping one for the trip back. It takes the last 1,000 bananas and drops 999 off at the 201st kilometer. All in all, the elephant ate three bananas to carry $998 + 999 = 1{,}997$ bananas from the 200th kilometer to the 201st kilometer. It continues and carries 1,994 bananas to the 202nd kilometer. It consumes three bananas for each kilometer to carry the rest and continues in this way until it reaches the 503rd kilometer with 1,001 bananas. It then leaves behind one banana and leaves directly for the oasis with the 1,000 bananas remaining. It arrives 467 kilometers later with 533 bananas.

**Proof of optimality** We refer to any path starting from the start and ending at the oasis as a strategy. As such, a strategy is one possible solution to the problem and we want to show that there is no better strategy than those that allow the merchant to bring 533 bananas to the oasis.

Let us begin by showing that if a strategy $(S)$ can allow a given number of bananas $n$ to be carried to the oasis, then there is another strategy $(S')$ that also allows $n$ bananas to be carried to the oasis but with a step in which the elephant is at the 999th kilometer with all the available bananas next to the elephant.

We consider a strategy $(S)$ that allows $n$ bananas to be carried to the oasis. In this strategy, the elephant will stop a given number of times, $p$, at the oasis to carry the bananas. At pass $i \in \{1, ..., p\}$, the elephant drops off a number $b_i$ of bananas at the oasis.

We can then construct a strategy $(S')$ from strategy $(S)$. The elephant performs the same actions as in strategy $(S)$, but instead of dropping the bananas off directly at the oasis, the elephant will stop 1 km before (the 999th kilometer) and drop off $b_i + 2$ bananas there (instead of dropping $b_i$ at the oasis and consuming two for the round trip). The elephant then continues as in strategy $(S)$ since it has the same number of bananas on its back as if it had done the round trip to the oasis. Before the last pass at the oasis in strategy $(S)$, the elephant is at the 999th kilometer with $b_p + 1 + \sum_{i=1}^{p-1} (b_i + 2)$ bananas by its side. Now the elephant carries all these bananas to the oasis in $p$ trips with a packet of $b_i$ bananas each time. We thus have a strategy $(S')$ in which the elephant can carry the same number of bananas but with a step where the elephant is at the 999th kilometer with all the available bananas next to it.

Finding the best strategy $(S)$ to carry 3,000 bananas to an oasis 1,000 km away is thus the same as finding the best strategy $(S')$ to carry 3,000 bananas to an oasis 999 km away. We deduce by induction that the best strategy is thus to carry all the bananas to the first kilometer, then all those remaining to the second kilometer, and so on until the oasis. This strategy is the one in which the elephant moves the entire cargo kilometer by kilometer. It allows the merchant to carry 533 bananas to the oasis. We can't do any better!

**70** Let us color the small cubes in black and white in such a way that each small cube is surrounded by adjacent cubes of the other color, as below:

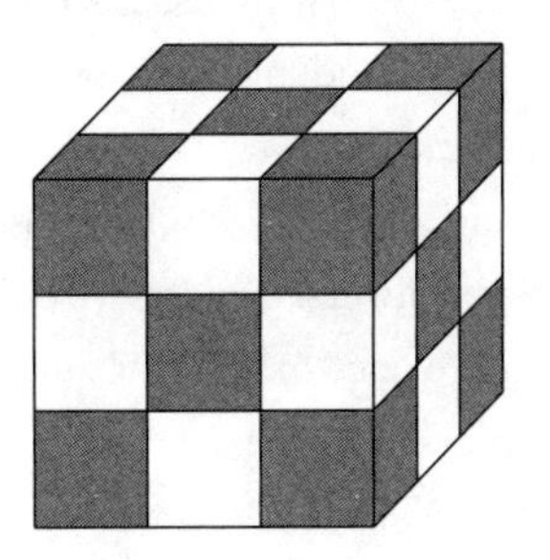

The termite travels through cubes of alternating colors. Since it eats an odd number of cubes, it must begin with a white cube to finish with the central cube, which is white. But then, if the termite eats all the cubes starting with a white cube and ending with a white cube, there will be one more white cube than black cubes. But there are 13 white cubes and 14 black cubes. It is thus impossible.

**71** Let us begin by solving the problem in the instance in which the students know that the lightbulb begins turned off. The students designate a leader who will be in charge of counting the calls to the office and announcing when everyone has been in. Other than the leader, each student flips the light on if it is off, but only the first time they're called and find it turned off. In other words, if a student finds the light off when they enter, but they have already turned it on during a previous visit, they will leave it off. When the leader is called, she always turns off the light if it was turned on. Thus, when the leader has turned the light off 99 times, it will be because the 99 other students have been in the office and the leader will be able to announce it.

To handle the general case, we slightly tweak the previous strategy. Other than the leader, the students turn the light on if it is turned off, twice but no more. If the light is initially turned off, the leader will have to turn it off $99 \times 2 = 198$ times; otherwise the leader will have to turn it off once more, that is, 199 times. In any case, the leader will know everyone has been in the office at least once when she turns the light off for the $198^{\text{th}}$ time.

**72** The map from the number of a student to the number of a drawer containing that student's number is, by assumption, a permutation that we will note as $\sigma$ (see reminder 4). Students may agree on the following strategy: each student first tries the drawer of their own number, then the drawer of the number they find inside, and so on until they find their number. In other words, student $i$ tries drawers $i$, $\sigma(i)$, $\sigma(\sigma(i))$, and so on.

We can decompose $\sigma$ into cycles of disjoint supports. The strategy works if and only if $\sigma$ contains no cycles of length strictly greater than 50. Indeed, if there is such a cycle, all the students whose numbers belong to this cycle will not find their numbers (they would find it after trying more than 50 drawers). If there is no cycle of length strictly greater than 50, all students have a number belonging to a cycle of length 50, so the strategy works. It remains to compute the probability $p$ that $\sigma$ verifies this property.

Let $N$ be the number of permutation of $\{1, ..., 100\}$ containing a cycle of length strictly greater than 50. Let $k \in \{51, ..., 100\}$; to choose a permutation of $\{1, ..., 100\}$ containing a cycle of length $k$ is to choose a support of this cycle among the $\binom{100}{k}$ possible supports, then the cycle among the $(k-1)!$ possible cycles of length $k$, and finally the rest of the permutation among the $(100-k)!$ possibilities. Thus, there are $\binom{100}{k}(k-1)!\,(100-k)! = \frac{100!}{k}$ permutations of $\{1, ..., 100\}$ containing a cycle of length $k$. Every permutation of $\{1, ..., 100\}$ whose largest cycle is of length strictly greater than 50 contains only one such cycle; therefore

$$N = \sum_{k=51}^{100} \frac{100!}{k}$$

The probability that the strategy fails is thus

$$1 - p = \frac{N}{100!} = \sum_{k=51}^{100} \frac{1}{k}$$

Hence,

$$p = 1 - \sum_{k=51}^{100} \frac{1}{k}$$

Numerically, $p \simeq 0.312$: the strategy does indeed have a success probability greater than 30%.

What if there were more than 100 students? With $2n$ students and the rule "a student may not open more than $n$ drawers," we obtain a probability of success

$$p_n = 1 - \sum_{k=n+1}^{2n} \frac{1}{k}$$

We notice that $\lim_{n \to +\infty} p_n = 1 - \ln(2)$, which is still greater than 30%.

**73** We begin by numbering the seven colors from 0 to 6. If a student knows the sum of the colors of the seven hats mod 7 (see reminder 6), they may deduce from it the color of their own hat. But the students have no way to know what the sum mod 7 of these colors will be. All they know is that this sum, mod 7, will be between 0 and 6. Nonetheless, the students can agree that the first student behaves as if the sum were congruent to 0 mod 7, and announces the color of their hat in consequence, the second student behaves as if the sum were congruent to 1 mod 7, the seventh, to 6 mod 7. In this way, exactly one student will guess their own color.

**74** Let us begin by presenting a possible strategy and then show that it answers the problem. First of all, the students number themselves from 1 to 7. Let us assume the perspective of student number 1. She observes the hats of the six other students and orders them in increasing order of their hats. Let us suppose she obtains the following order $5, 2, 4, 7, 3, 6$. She then adds her own number to the end of the ranking: $5, 2, 4, 7, 3, 6, 1$. This is a permutation of $\{1, 2, 3, 4, 5, 6, 7\}$ (see reminder 4). She now only needs to compute the signature of this permutation and to take a black T-shirt if the signature is $-1$ and a white T-shirt if the signature is 1.

Let us now see why this strategy works. First of all, one must be convinced of the following property: "If the adopted strategy is such that whenever two students have on their hats neighboring real numbers, they choose different colored T-shirts, then the students will indeed have alternating colors."

Next, one must show that the given strategy upholds this property. To see this, consider an example:

| students | 1 | 2 | 3 | 4 | 5 | 6 | 7 |
|---|---|---|---|---|---|---|---|
| real numbers | $-\pi$ | 0 | $\sqrt{2}$ | $\frac{5}{11}$ | $-0.1$ | 1,000 | $\frac{1}{2}$ |

In this example, students are sorted into order $1, 5, 2, 4, 7, 3, 6$. We notice that students 4 and 7 have neighboring real numbers (but they do not know it). Let us assume the perspective of student 4 and apply the previously designed strategy. Student 4 sees the six other students in the order $1, 5, 2, 7, 3, 6$. He adds his number on the end, and computes the signature of the permutation mapping $1, \ldots, 7$ to $1, 5, 2, 7, 3, 6, 4$. Now we consider student 7. She sees $1, 5, 2, 4, 3, 6$ and computes the signature of the permutation mapping $1, \ldots, 7$ to $1, 5, 2, 4, 3, 6, 7$. In fact, we notice that since students 4 and 7 are neighbors, they will see the same order for the six other remaining students except that a 4 will become a 7 and a 7 will become a 4. They will thus compute the signature of their permutation up to one transposition: going from $1, 5, 2, 7, 3, 6, 4$ to $1, 5, 2, 4, 3, 6, 7$ by transposition of the 4 and the 7. The signature of the first permutation will thus be the opposite of the signature of the second. Students 4 and 7 will thus choose T-shirts of different colors. Thus, the property holds, and the presented strategy works.

**75** Let us first solve the simple instance. Start by filling the 3 liter pitcher with wine from the barrel. Now fill the 5 liter pitcher with wine from the 3 liter pitcher. We fill up the 3 liter pitcher again and pour the wine from that into the 5 liter pitcher until it is full. The remainder in the 3 liter pitcher is exactly 1 liter.

Let us now move to the general case. The integers $p$ and $q$ are coprime, so by Bézout's lemma, there are two integers $u, v \in \mathbb{Z}$ such that $up + vq = 1$. Since $p$ and $q$ are positive, either $u$ or $v$ is negative. Let us assume that $v$ is negative (the case for $u$ negative follows in the same manner). We can then write $up = 1 + |v|q$. It suffices to fill the $q$ liter pitcher with $up$ liters obtained by filling the $p$ liter pitcher $u$ times and emptying it into the other. Indeed, if we note $r$ as the number of liters in the $p$ liter pitcher after the procedure, we have $up \equiv r\,[q]$, but $up \equiv 1\,[q]$, thus $r \equiv 1\,[q]$. Since $0 \leq r < q$, $r = 1$.

For example, integers 11 and 47 are coprime. We thus have the following Bézout relation: $4 \times 47 - 17 \times 11 = 1$; hence, $4 \times 47 = 1 + 17 \times 11$. Thus, to obtain 1 liter, it is sufficient to fill the 47 liter pitcher four times while pouring the wine into the 11 liter pitcher while emptying the 11 liter pitcher each time it is full. The 11 liter pitcher will be filled 17 times; at this moment, 1 liter will remain in the 47 liter pitcher. When the first 47 liter pitcher is empty, there are 3 liters left in the 11 liter pitcher. Now fill the 47 liter pitcher again and repeat the process. After this, there are now 6 liters in the 11 liter pitcher. Repeating this procedure again leaves 9 liters in the 11 liter pitcher, and the fourth repetition will push it past 11 liters so that it needs to be emptied once more, and then it will contain exactly 1 liter at the end.

**76** Let us begin with the case in which $q = 2$. Let us show that with $n$ rats, over the course of 2 days, we cannot discriminate between more than $3^n$ bottles; we will then show how to discriminate between $3^n$ bottles. Suppose we can discriminate between $k$ bottles. Once more, we compare the number of outcomes and hypotheses. There are $k$ bottles and hence $k$ possible solutions to the problem. Let us count the number of outcomes by counting the number of dead rats at the end of the first day. If no rats died the first day, there are $n$ rats left, so there are $2^n$ possible outcomes for the second day. If only one rat died the first day, there are $n-1$ rats and hence $2^{n-1}$ outcomes possible on the second day. If $i$ rats died the first day, there are $n-i$ rats left and hence $2^{n-i}$ possible outcomes for the second day. If all the rats died the first day, there are none left, so the only outcome left for the second day is that all rats have died.

How many cases are there exactly if $i$ rats die the first day? The answer is $\binom{n}{i}$. Thus, the number of possible outcomes is the following:

$$\sum_{i=0}^{n} \binom{n}{i} \times 2^{n-i} = 3^n$$

By comparing the number of possible outcomes to the number of possible solutions to the problem, we obtain $k \leq 3^n$, which is what we claimed.

Let us show how we can discriminate between $3^n$ bottles with $n$ rats in 2 days. We begin by dividing the bottles into $n+1$ groups numbered 0 to $n$. The first group (number $k = 0$) contains $2^n$ bottles. The second group contains $n \times 2^{n-1}$ bottles spread out into $n$ subgroups of $2^{n-1}$ bottles. Group number $k$ contains $\binom{n}{k} \times 2^{n-k}$ bottles spread into $\binom{n}{k}$ subgroups of $2^{n-k}$ bottles. All the bottles are accounted for because

$$\sum_{k=0}^{n} \binom{n}{k} \times 2^{n-k} = 3^n$$

No rat tastes any bottle of the first group. For the second group, each rat tastes all the bottles of a subgroup: we have $n$ rats, and there are $n$

subgroups. For group $k$, each possible group of $k$ rats tastes all the bottles of a subgroup: there are $\binom{n}{k}$ groups of $k$ rats and there are $\binom{n}{k}$ subgroups. The next day, if $k$ rats died, we know the poisoned bottle is in group number $k$, and by looking at which group of $k$ rats died, we also know to which subgroup the poisoned bottle belongs. This subgroup contains $2^{n-k}$ bottles and we have $n-k$ rats. We can thus find the poisoned bottle in 1 more day by applying the method found in the 1 day version (riddle 57).

By induction, we show that $n$ rats can discriminate $(q+1)^n$ bottles over $q$ days, and that it isn't possible to do better. It suffices to use the same method by exploiting the following relation:

$$\sum_{i=0}^{n} \binom{n}{i} \times q^{n-i} = (q+1)^n$$

To conclude, if there are $p$ bottles and we have $q$ days, it is necessary and sufficient that the number $n$ of rats satisfies the requirement that $(q+1)^n \geq p$. If one wants a tight formula, the minimal number of rats is

$$n = \left\lceil \frac{\ln(p)}{\ln(q+1)} \right\rceil$$

where $\lceil . \rceil$ is the ceiling function.

**77** Everything boils down to the fact that $\mathbb{Z} \times \mathbb{Z}$ is countable, which means we have (see reminder 4) a bijection

$$\sigma : \begin{cases} \mathbb{N} & \to \mathbb{Z} \times \mathbb{Z} \\ k & \mapsto (\sigma_1(k), \sigma_2(k)) \end{cases}$$

If we know $(m, p)$, we know where the robot is after $k$ seconds: it is at $m + kp$. We can test the couples $(m, p)$ by asking at each instant $k$ if the robot is at $\sigma_1(k) + k \times \sigma_2(k)$.

Let us check if this strategy works; $\sigma$ is a bijection, so there exists $n \in \mathbb{N}$ such that $\sigma(n) = (m, p)$. At the $n$th second, we will ask if the robot is in $m + np$ and it will indeed be there. We notice that the robot can be found before the $n$th second and that when we find it at instant $k$, nothing shows that $(m, p) = \sigma(k)$. Thus, we cannot determine its starting position, nor its step size, but the main thing is that we've found it.

**78** We can convince ourselves that there are no solutions by trying different routes. This is what we shall prove. To do so, we will need to use a bit of graph theory, created especially for this problem by Euler. This approach is very clever, yet very simple. The idea is to reduce the problem to connecting dots without lifting the pen off the paper. We give each part of town a number as follows:

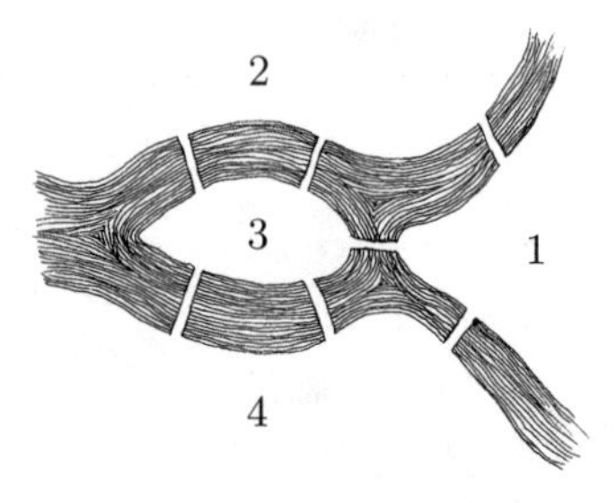


If there is a bridge between two regions, we connect those regions with a line or a curve; the result is the following graph:

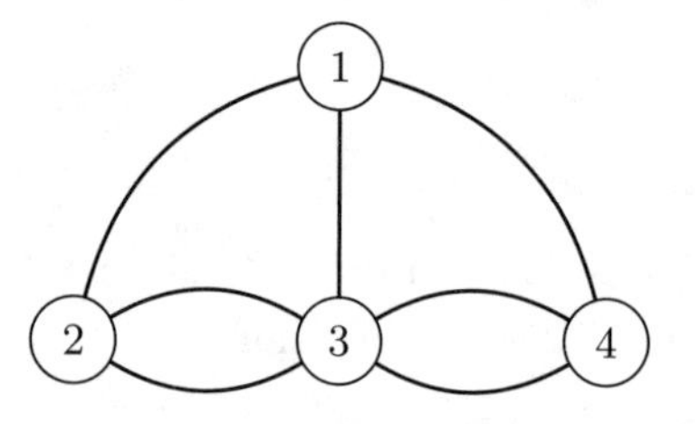


We then notice that finding a path that crosses each bridge exactly once boils down to finding a path linking the four vertices of the graph without lifting the pen up. One is easily convinced by trying that it is impossible, but it remains to prove it. To do so, we will use the following property:

*If a graph can be drawn without lifting the pen, then it has at most two vertices with an odd number of edges.*

Consider as an example the following graph:

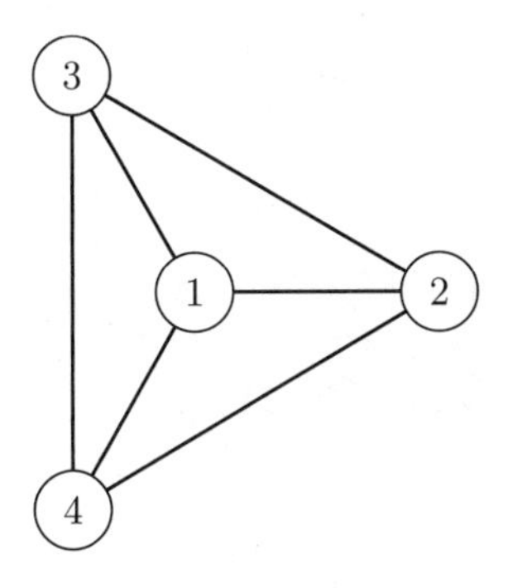


On each of the four vertices, there are three edges: we say that each vertex is of degree three. Since $4 > 2$, it is impossible to draw this graph without lifting the pen.

Let us now prove the result. Consider a graph that we assume to be drawn without lifting the pen. It has a vertex from which we started drawing and a vertex where we finished drawing; the two vertices may be the same. We notice that all other vertices must be of even degree. Indeed, each time we draw through a vertex, we increase its degree by two, which makes the final degree even. Thus, there are indeed at most two vertices of odd degree.

Let us return to the bridge problem. All the vertices of the graph have an odd degree (3 for vertices 1, 2, and 4; 5 for vertex 3). It is thus impossible to draw without lifting the pen.

Do note that the converse of this result is true but it is somewhat more difficult to prove.

**79** Once more, we must reduce to a graph problem. We begin by numbering each part of the plane separated by a line, as below:

1

2 3 4

5 6

We connect two regions by an edge when they are separated by a line in the original problem; doing so yields the following graph:

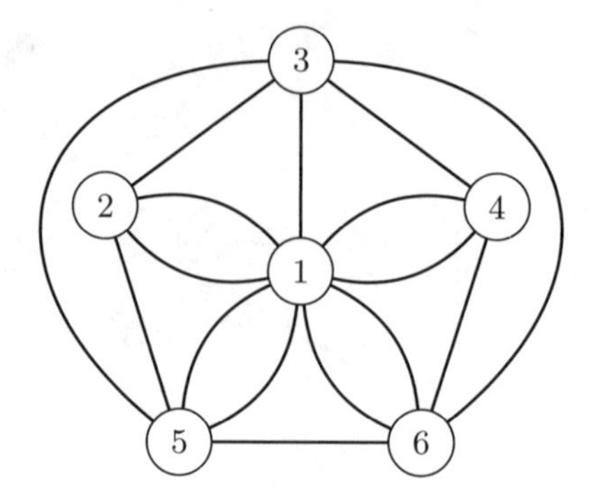


Drawing a line that passes through each edge exactly once in the problem is the same as finding a way to connect the six edges of the graph by passing over its edges without lifting the pen. On vertices 1, 3, 5, and 6, the number of edges is odd. In the solution to the previous riddle, we learned that we cannot link the vertices of a graph without lifting the pen off the page if there are more than two vertices of odd degree. Here, we have four odd vertices. Thus, it is impossible to connect the edges of the graph by passing over its edges without lifting the pen.

**80** Cards are numbered from 0 to 123. A member of the audience chooses five cards. Let $c_0 < c_1 < c_2 < c_3 < c_4$ be their numbers. Bob chooses card $c_k$, with $k \in \{0, ..., 4\}$ such that $k \equiv c_0 + c_1 + c_2 + c_3 + c_4\,[5]$. Let $s$ be the sum mod 5 of the four remaining cards. $k \equiv s + c_k\,[5]$, thus $c_k \equiv k - s\,[5]$. Alice, learning the four cards, can renumber the 120 other cards from 0 to 119 while preserving the order. Thus, the card chosen by Bob is no longer congruent to $k - s$ mod 5 but to $-s$ mod 5. Indeed, $k$ cards were removed, so the new number of cards is the old number of cards minus $k$. There are exactly 24 cards congruent to $-s$ mod 5 among the 120. Since there are 24 ways to order the remaining 4 cards, Bob was able to communicate the remaining information (a number between 1 and 24) to Alice, which allows her to conclude which card Bob chose.

Let us give an example to clarify this method. Let us assume that the member of the audience chose card numbers 92, 118, 110, 95, and 8.

**Encoding** Bob starts by ordering the cards. He computes their sum mod 5: $3 + 2 + 0 + 0 + 3 = 8$; thus their sum is congruent to 3 mod 5. He thus chooses card 110. The sum of the remaining cards mod 5 is 3. By renumbering the 120 cards, card 110 becomes card 107. $107 = 21 \times 5 + 2$. Bob uses the four remaining cards to encode 21 by putting them back in the order 95, 117, 8, 92.

**Decoding** Alice looks at the four cards. By looking at the order the cards are in, she deduces that Bob has encoded 21. Moreover, their sum is equal to 3 mod 5, so the number of the mystery card in the new numbering (from 0 to 119) is congruent to 2 mod 5. Since Bob coded 21, it is equal to $21 \times 5 + 2 = 107$. Three cards among the four are below 107; thus the real number is 110.

**81** It is impossible. Let us assume for contradiction that we have successfully paved the bathroom. We then paint the tiles as follows:

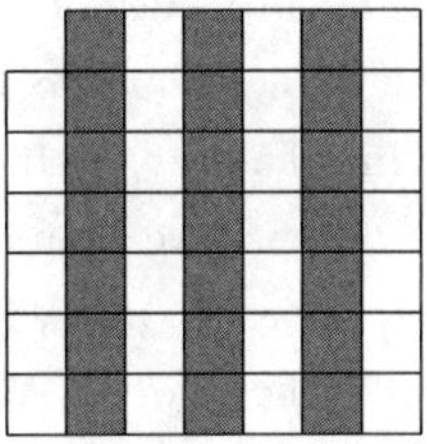

We notice that there are 27 white tiles and 21 gray tiles. Each L-shaped tile covers three white tiles and one gray or the other way around. Let $W$ be the number of L-shaped tiles that cover three white tiles and one gray and $G$ the number of L-shaped tiles that cover three gray tiles and one white. We obtain the following system:

$$\begin{cases} 3W + G &= 27 \\ W + 3G &= 21 \end{cases}$$

Solving this system, the only solution is $W = \frac{15}{2}$ and $G = \frac{9}{2}$. This result is a contradiction since $W$ and $G$ must be whole numbers.

**82** It is impossible. Let us assume for contradiction that we have successfully paved the bathroom. We then paint the tiles as follows:

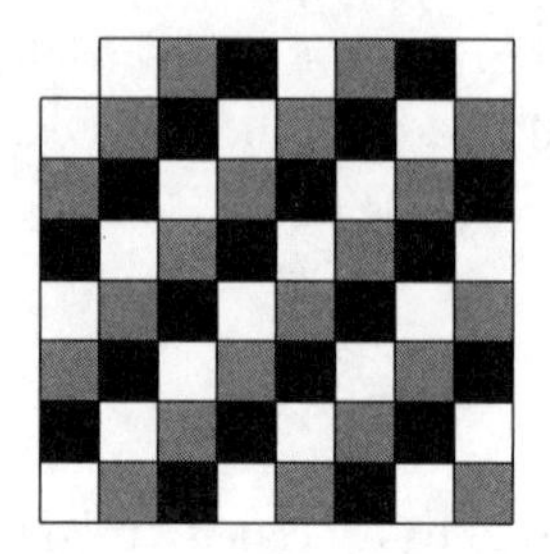

We notice that there are 22 white tiles, 21 gray tiles, and 20 black tiles. On the other hand, each domino-shaped tile must cover a white square, a gray square, and a black square. Thus, the numbers of white tiles, gray tiles, and black tiles must all be equal to each other. This result is a contradiction.

**83** It is house 73 that will get its mail last. Let us compute generally for $n$ houses ($n \geq 2$) which one gets its mail last. Let $m_n$ be the number of this house. By observing the small values of $n$ ($m_2 = 1$, $m_3 = 3$, $m_4 = 1$, $m_5 = 3$,...) we can notice that the sequence will increase by taking only odd values until $n$ before falling back to 1. It is not easy to write this as an explicit formula. This is because we aren't working in the right base. Let us consider base 2 (see reminder 7); $n$ is written as $1b_{p-1}...b_0$ in base 2, with $p \geq 1$ and $b_i$ taking value either 0 or 1. We will show, by induction on $n \geq 2$, that in base 2, $m_n$ is written as $b_{p-1}...b_0 1$.

**Basis** For $n = 2 : 2 = 10_2$ ($10_2$ means 10 in base 2), $m_n = 1 = 01_2$. This is also true for $n = 3 : 3 = 11_2$ and $m_3 = 3 = 11_2$.

**Induction** Assume the property holds for some rank $n \geq 3$ and show it holds at rank $n+1$. The postman delivers mail to house number 2. He must then deliver the mail to the $n$ remaining houses. We can use the induction hypothesis if we renumber the houses beforehand as follows:

| 1 | 2 | 3 | 4 | $\cdots$ | $n$ | $n+1$ |
|---|---|---|---|---|---|---|
| n | $\times$ | 1 | 2 | $\cdots$ | $n-2$ | $n-1$ |

House number 3 becomes house number 1, number 4 becomes number 2, ..., $n+1$ becomes $n-1$, and 1 becomes $n$.

If $m_n = n$, given the renumbering performed, $m_{n+1} = 1$. By uniqueness of the base 2 representation $b_0 = 1$, $b_1 = b_0$, ..., $b_{p-1} = b_{p-2}$, and thus $n = 1 \dots 1_2$. In this case, $n+1 = 10 \dots 0_2$, and the induction property holds: $m_{n+1} = 1$.

If not, $m_n \neq n$, and, given the numbering, $m_{n+1} = m_n + 2$.

$$n = 2^p + \sum_{i=0}^{p-1} b_i 2^i$$

$n+1$ has no more digits in base 2 than $n$, otherwise $n = 1...1_2 = m_n$, which is forbidden. Therefore,

$$n + 1 = 2^p + \sum_{i=0}^{p-1} B_i 2^i$$

with the $B_i$ taking values in $\{0, 1\}$.

$$B_{p-1}...B_0 1_2 = \sum_{i=0}^{p-1} B_i 2^{i+1} + 1$$

$$B_{p-1}...B_0 1_2 = 2\left(\sum_{i=0}^{p-1} b_i 2^i + 1\right) + 1$$

$$B_{p-1}...B_0 1_2 = \left(2\sum_{i=0}^{p-1} b_i 2^i + 1\right) + 2$$

Thus $B_{p-1}...B_0 1_2 = b_{p-1}...b_0 1_2 + 2 = m_n + 2 = m_{n+1}$. This calculation completes the induction. $100 = 1100100_2$, hence $m_{100} = 1001001_2 = 73$ as claimed.

**84** Let us start by establishing a strategy that is not optimal but is already better than the one proposed in the statement. Let us denote $R$ as the radius of the lake, $v$ as the speed of the dog, and $v'$ as the speed of the tourist. Let us consider the disk or radius $r = \frac{v'}{v} R$. As long as the tourist remains inside this disk, her maximum angular speed is faster than the dog's. In other words, as long as the tourist is moving in a circle whose radius is less than $r$, she moves around the center faster than the dog. By reasoning in the limit case, she can, after a given number of cycles, place herself on the opposite side from the dog on the circle of radius $r$. Now the tourist will swim straight for the shore:

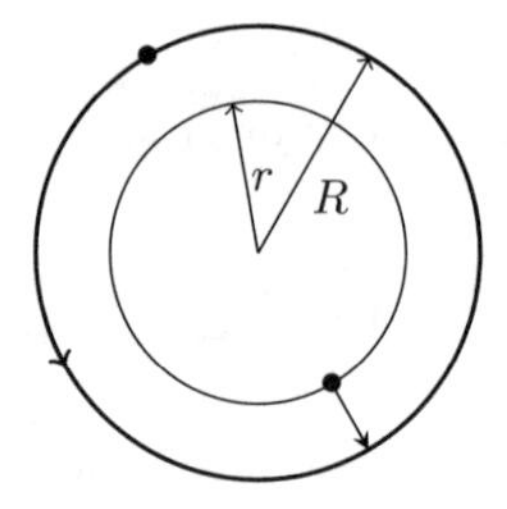


She travels a distance of $R - r = R\left(1 - \frac{v'}{v}\right)$, which takes her

$$t' = \frac{R(1 - \frac{v'}{v})}{v'} = R\left(\frac{1}{v'} - \frac{1}{v}\right)$$

while the dog travels the distance $\pi R$, which takes it $t = \dfrac{\pi R}{v}$.
For the tourist to escape the dog, we must have $t' < t$; hence a limit speed of $v' = \dfrac{v}{1+\pi}$. As announced, this speed is lower than the one proposed in the statement, $\dfrac{v}{\pi}$, which makes this strategy better.

Let us now shift to the optimal strategy. As before, the tourist places herself on the circle of radius $r$ with the dog on the other side, but instead of escaping straight ahead, she follows a tangent to the circle:

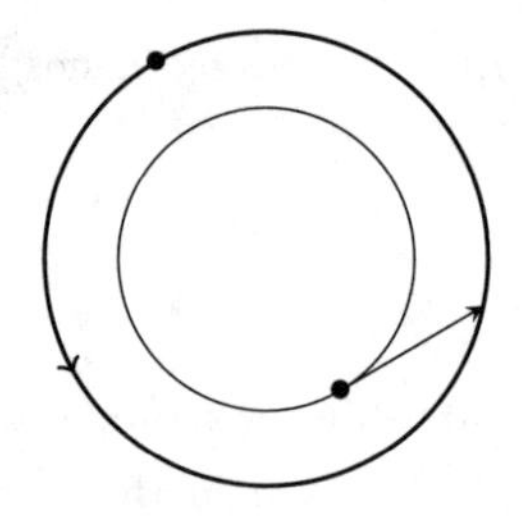

By reasoning in the limit case, we can assume that the dog leaves in the wrong direction, like on the drawing. Indeed, the tourist can leave straight for the shore (like in the previous strategy) for a very short interval, until the dog starts in one direction; then the tourist can take the tangent in the other direction:

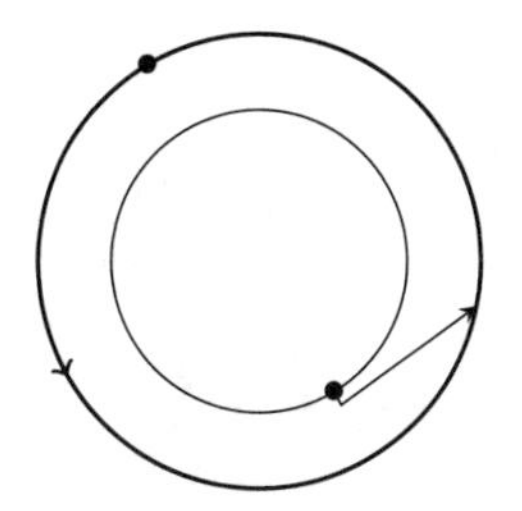

As soon as the tourist leaves the disk of radius $r$, the dog's angular speed is always greater, so the dog can't turn around. She travels a distance of $\sqrt{R^2 - r^2}$, which takes her

$$t' = \frac{\sqrt{R^2 - r^2}}{v'}$$

while the dog travels $\pi\,(R + \alpha)$, which takes it

$$t = \frac{\pi\,(R + \alpha)}{v}$$

So that the tourist escapes the dog, we must have $t' < t$. By setting $x = \frac{r}{R} = \frac{v'}{v}$, we obtain after some transformations the following condition:

$$\left(1 + (\pi + \arccos(x))^2\right) x^2 = 1$$

We can't obtain an exact expression for $x$, but by plugging the computation into a suitable computer solver, we can obtain $x \simeq 0.217$. This is better than the previous strategy, since $\frac{1}{1+\pi} \simeq 0.241$. It remains to show that this is optimal.

This proof can be done without calculation. Let us first prove that the strategy of taking the tangent is better than swimming perpendicular to the circle (straight for shore). If the tourist leaves from point $A$ to go to point $B$ perpendicular to the circle, like in the first strategy, we can assume she in fact leaves from point $A'$, goes to point $A$ and then on to point $B$. This changes nothing; the dog stays on the opposite side of the pond while the tourist goes from $A$ to $A'$. But to go from point $A'$ to point $B$, the shortest distance is the straight line, that is, the tangent. The tangent method thus allows the tourist to leave faster.

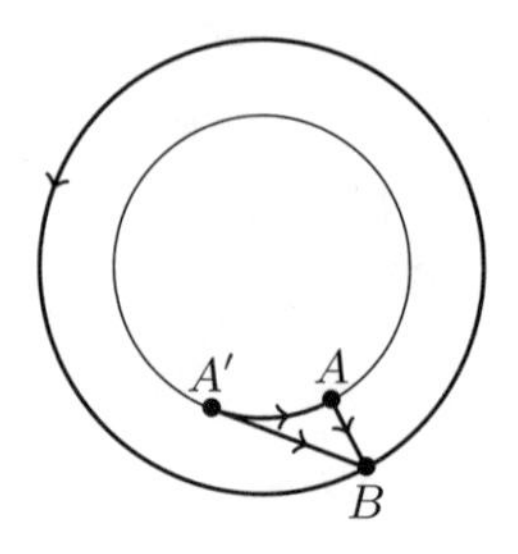


Let us now take an arbitrary strategy $(S)$ allowing the tourist to escape and let us show that the strategy $(T)$ of taking the tangent is better. With $(S)$, the tourist leaves the disk of radius $r$ from a point $A$ and arrives in point $B$. Up to improving $(S)$, we can always assume that when the tourist is in $A$, the dog is on the opposite side of the lake. By symmetry, we can always assume that the dog runs counterclockwise. By introducing $B'$, the opposite of $B$ on the circle of radius $r$, we can separate two cases:

- If $A$ is in the right part of the circle, between $A'$ and $B'$, we apply the same thinking as before: we can assume that in fact $(S)$ starts from $A'$.

- If $A$ is not in the right part of the circle, we can assume that in fact $(T)$ starts from $A$. This doesn't change $(T)$ but allows us to compare it to $(S)$.

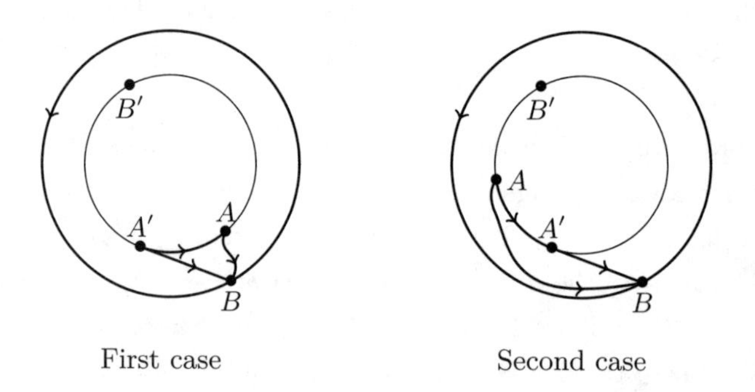


First case Second case

In either case, we see that the strategy $(T)$ has a shorter exit time than strategy $(S)$: therefore, $(T)$ is the best strategy.

**85** A natural strategy is to fix a threshold above which one is happy, for example $100, and to cash the first check if it is greater than $100, but otherwise to cash the other check.

If both sums $M_1$ and $M_2$ chosen by the banker are below $100, this strategy leads the client to always change envelopes. He will thus gain on average $\frac{M_1+M_2}{2}$. If both sums chosen by the banker are above $100, this strategy leads the client to never change envelopes. Thus, he wins, on average, $\frac{M_1+M_2}{2}$. Finally, if $M_1 < 100 < M_2$ or the other way around, this strategy allows the client to always choose the envelope with the highest amount.

In the end, the strategy consisting of thresholding the amount one is happy with is only a good strategy if the threshold is between the two sums the banker chose. If we fix it arbitrarily, it is possible that it doesn't land between the two checks and that the strategy is ineffective. To avoid this issue, we must choose the threshold at random, with a random variable $X$ such that

$$\forall (a,b) \in (\mathbb{R}_+)^2, a < b \Rightarrow \mathbb{P}(X \in [a,b]) > 0$$

In other words, $X$ is a random variable with a nonzero probability of landing in any interval of $\mathbb{R}_+$, such as an exponential law.

In this manner, the client chooses the threshold at random with the random variable $X$, and he knows that sometimes, the chosen threshold will be in between the two sums chosen by the banker. In any case, he performs better than $\frac{M_1+M_2}{2}$. His average gain is thus strictly better than $\frac{M_1+M_2}{2}$. We can compute the average gain $G$ as a function of $M_1$ and $M_2$ and of $p = \mathbb{P}(X \in [M_1, M_2])$

$$G = p \times \max(M_1, M_2) + (1-p) \times \left(\frac{M_1 + M_2}{2}\right)$$

**86** Let us take up the banker's reasoning: "Let $x$ be the amount of the left envelope (we don't know $x$ since it hasn't been opened). The sum in the envelope on the right has a 50/50 chance of being $2x$, and a 50/50 chance of being $\frac{x}{2}$; thus, the envelope on the right contains on average

$$\frac{1}{2} \times 2x + \frac{1}{2} \times \frac{x}{2} = \frac{5}{4}x$$

which is strictly more than $x$. Thus it is beneficial to change envelopes!"

The mistake is that the sum in the right envelope need not have a 50/50 chance of being $2x$, nor a 50/50 chance of being $\frac{x}{2}$. Even though there is always a certain amount in one envelope and twice as much in the other, one must do conditional probability: conditionally in this case to "the first envelope contains amount $x$" and this event changes the probability of having $2x$ and $\frac{x}{2}$ in the other envelope.

Let us take a simple example: assume the banker chooses the first sum at random between 1 and 100, and the second sum is twice the first, but that the two sums are assigned to either envelope at random. The probability that the second envelope contains $2x$ if the first contains $x = 200$ is 0.

Let us show that there can be no discrete probability law such that the sum in the right envelope always has a 50/50 chance of being $2x$ and a 50/50 chance of being $\frac{x}{2}$, regardless of what the initial sum $x$ of the left envelope was.

Let $X$ be a discrete random variable taking values $x_1, x_2, ..., x_n, ...$, respectively, with probabilities $p_1, p_2, ..., p_n, ...$, such that:

$$\forall k \in \mathbb{N}^*, p_k \in [0,1] \quad \text{and} \quad \sum_{k=1}^{+\infty} p_k = 1$$

Let $\epsilon$ be a random variable, independent of $X$ with value in $\{1, 2\}$ and such that

$$\mathbb{P}(\epsilon = 1) = \mathbb{P}(\epsilon = 2) = \frac{1}{2}$$

We model the sums of envelopes 1 and 2 by the random variables $E_1$ and $E_2$, defined by

$$E_1 = \epsilon \times X \quad \text{and} \quad E_2 = (3 - \epsilon) \times X$$

Thus, when $\epsilon = 1$, $E_1 = X$ and $E_2 = 2X$ and when $\epsilon = 2$, $E_1 = 2X$ and $E_2 = X$. According to the banker, the probability that the sum in the right envelope $E_2$ is equal to $2x$ given that the sum in the left envelope $E_1$ is $x$ is equal to the probability that the sum in the right envelope is equal to $\frac{x}{2}$ given that sum in the left envelope is $x$, he thus assumes that:

$$\mathbb{P}(E_2 = 2x | E_1 = x) = \mathbb{P}\left(E_2 = \frac{x}{2} | E_1 = x\right)$$

and thus

$$\frac{\mathbb{P}(E_2 = 2x, E_1 = x)}{\mathbb{P}(E_1 = x)} = \frac{\mathbb{P}\left(E_2 = \frac{x}{2}, E_1 = x\right)}{\mathbb{P}(E_1 = x)}$$

$$\mathbb{P}(E_2 = 2x, E_1 = x) = \mathbb{P}\left(E_2 = \frac{x}{2}, E_1 = x\right)$$

But

$$\mathbb{P}(E_2 = 2x, E_1 = x) = \mathbb{P}(\epsilon = 1, X = x)$$

and the random variables $\epsilon$ and $X$ are independent; thus,

$$\mathbb{P}(\epsilon = 1, X = x) = \mathbb{P}(\epsilon = 1) \times \mathbb{P}(X = x) = \frac{1}{2} \times \mathbb{P}(X = x)$$

Likewise,

$$\mathbb{P}\left(E_2 = \frac{x}{2}, E_1 = x\right) = \frac{1}{2} \times \mathbb{P}\left(X = \frac{x}{2}\right)$$

and thus

$$\frac{1}{2} \times \mathbb{P}(X = x) = \frac{1}{2} \times \mathbb{P}(X = 2x)$$

which contradicts

$$\sum_{k=1}^{+\infty} p_k = 1$$

**87** There are two ways to proceed. Either we take the result already established for whole numbers and generalize to arbitrary real valued weights, or we reason directly on arbitrary weights with some matrix algebra. Both these methods require the use of vector spaces.

**First method** Let us begin by assuming that the weights are rational. First, we reduce first to positive rational numbers by adding a sufficiently large number to all weights, then by multiplying all the weights by the product of denominators; in this way, we can reduce to the case of natural numbers, already solved in riddle 63.

Passing to real weights is much more subtle. The idea is to consider $\mathbb{R}$ as a $\mathbb{Q}$ vector space. The family of 11 weights $(x_1, \ldots, x_{11})$ becomes a family of 11 vectors. Let $E$ be the vector space spanned by these 11 vectors, and let $(e_1, \ldots, e_n)$ be a basis for $E$. (We don't know the dimension $n$ of $E$, we only know that $n \leq 11$). For all $i \in \{1, \ldots, 11\}$, there are thus $\lambda_{i,1}, \ldots, \lambda_{i,n} \in \mathbb{Q}$ such that $x_i = \lambda_{i,1}e_1 + \cdots + \lambda_{i,n}e_n$.

Since $(e_1, \ldots, e_n)$ are linearly independent, each family $(\lambda_{i,j})_{1 \leq i \leq 11}$ is a distribution of weights verifying the property noticed by the squirrel. Since all these weights are rational, we can conclude they are all equal and that this is true for any $j$ between 1 and $n$. Thus, all the $x_i$ are equal.

**Second method** Consider a distribution that works. Note $X$ as the vector of $\mathbb{R}^{11}$ representing the weights of the nuts. The condition on these weights is $MX = 0$ with $M \in \mathcal{M}_{11}(\mathbb{R})$, a matrix with zeroes on the diagonal and five 1s and five $-1$s on each line. Let $H$ be the vector in $\mathbb{R}^{11}$ whose coordinates are only 1s. We note that $MH = 0$, so $H \in \ker(M)$. Let us show that $\ker(M)$ is of dimension 1. By the rank-nullity theorem, it is sufficient to show that $M$ is of rank 10. To do so, consider $M'$ the upper-left submatrix of size 10 of $M$.

$$M' = \begin{pmatrix} 0 & \pm 1 & \cdots & \pm 1 \\ \pm 1 & \ddots & \ddots & \vdots \\ \vdots & \ddots & \ddots & \pm 1 \\ \pm 1 & \cdots & \pm 1 & 0 \end{pmatrix}$$

It is sufficient to compute the determinant to check that it is nonzero. By the determinant formula and the compatibility of congruence with addition and multiplication,

$$\det(M') \equiv \det \begin{pmatrix} 0 & 1 & \cdots & 1 \\ 1 & \ddots & \ddots & \vdots \\ \vdots & \ddots & \ddots & 1 \\ 1 & \cdots & 1 & 0 \end{pmatrix} [2]$$

Let us denote $D$ as this last determinant. It is computed simply by combining the lines and columns. By subtracting the first column from the others,

$$D = \det \begin{pmatrix} 0 & 1 & \cdots & \cdots & 1 \\ 1 & -1 & 0 & \cdots & 0 \\ 1 & 0 & \ddots & \ddots & \vdots \\ \vdots & \vdots & \ddots & \ddots & 0 \\ 1 & 0 & \cdots & 0 & -1 \end{pmatrix}$$

Then, by adding all the other lines to the first line

$$\begin{aligned} D = \det & \begin{pmatrix} 9 & 0 & \cdots & \cdots & 0 \\ 1 & -1 & \ddots & & \vdots \\ 1 & 0 & \ddots & \ddots & \vdots \\ \vdots & \vdots & \ddots & \ddots & 0 \\ 1 & 0 & \cdots & 0 & -1 \end{pmatrix} \\ & = 9 \times (-1)^9 \end{aligned}$$

Thus, $D \equiv 1\,[2]$, so indeed the matrix is of rank 10 and $\ker(M)$ is of dimension 1. Since $X \in \ker(M)$, there is $\alpha \in \mathbb{R}$ such that $X = \alpha H$. Thus all the weights are equal.

**88** $t_1 > t_2$: On average we will have to wait longer for the monkey to type "abracadabra" than "abracadabrx." The two average times differ because when the monkey successfully types "abracadabr," if its job is to type "abracadabra," it either types in an "a" and succeeds, or it types another letter and must start over. On the other hand, if the monkey must type "abracadrabrx," it either types "x" and succeeds right away, or it fails but it doesn't necessarily start over from scratch. If the monkey typed in an "a" instead of an "x," it must start over to type "abracadabrx," but it starts over having already completed the "abra"! Without being rigorous, this reasoning justifies the conjecture that $t_1 > t_2$.

To compute $t_1$, we will proceed in an analogue manner to riddle 50. Let us define the following 11 steps:

| Step | The monkey has typed |
|---|---|
| 0 | xxx |
| 1 | xxxa |
| 2 | xxxab |
| 3 | xxxabr |
| 4 | xxxabra |
| 5 | xxxabrac |
| 6 | xxxabraca |
| 7 | xxxabracad |
| 8 | xxxabracada |
| 9 | xxxabracadab |
| 10 | xxxabracadabr |

Step 0 corresponds to the state where the monkey "starts from scratch," for example, if it typed "jhusdqabracadakdf." Step 1 corresponds to the state where the monkey starts from the letter "a," for example if it typed "jhusdqabracadakdfa." Let $T_0, \ldots, T_{10}$ be the average times to obtaining "abracadabra" from step $0, \ldots, 10$. We seek $T_0$, since $T_0 = t_1$.

When the monkey is at step 0, either it types "a" and passes to step 1 or it types any other letter and stays at step 0. If it types "a," its average remaining time becomes $T_1$; thus, its average (total) time is $T_1 + 1$, since it has typed one more letter. Otherwise, it retains an average time of $T_0$, and thus a total time of $T_0 + 1$. Thus

$$T_0 = \frac{1}{26}(T_1 + 1) + \frac{25}{26}(T_0 + 1)$$

Let us set $p = \frac{1}{26}$. By applying the same line of thought to each step, we finally obtain the following 11 relations:

$$\begin{aligned}
T_0 &= pT_1 + (1-p)T_0 + 1\\
T_1 &= pT_2 + pT_1 + (1-2p)T_0 + 1\\
T_2 &= pT_3 + pT_1 + (1-2p)T_0 + 1\\
T_3 &= pT_4 + (1-p)T_0 + 1\\
T_4 &= pT_5 + pT_2 + pT_1 + (1-3p)T_0 + 1\\
T_5 &= pT_6 + (1-p)T_0 + 1\\
T_6 &= pT_7 + pT_2 + pT_1 + (1-3p)T_0 + 1\\
T_7 &= pT_8 + (1-p)T_0 + 1\\
T_8 &= pT_9 + pT_1 + (1-2p)T_0 + 1\\
T_9 &= pT_{10} + pT_1 + (1-2p)T_0 + 1\\
T_{10} &= (1-p)T_0 + 1
\end{aligned}$$

We will not detail how to solve this linear system. The best thing to do is to plug it into solver software. The average time to obtain "abracadabra," the solution for $T_0$ above, works out to be

$$t_1 = 3{,}670{,}344{,}487{,}444{,}778\,\text{s}$$

To compute $t_2$, only the last of the relations is changed and becomes:

$$T_{10} = pT_4 + (1-2p)T_0 + 1$$

and we thus obtain

$$t_2 = 3{,}670{,}344{,}486{,}987{,}776\,\text{s}$$

$t_1 - t_2 = 457{,}002\,\text{s}$: so, indeed, $t_1 > t_2$.

**89** Let us consider a black and white tiling (pictured as gray and white for visibility) whose origin is the lower left corner of the large rectangle. Let us fix the tiles' shapes to be squares with edges of length $\frac{1}{2}$. This gives:

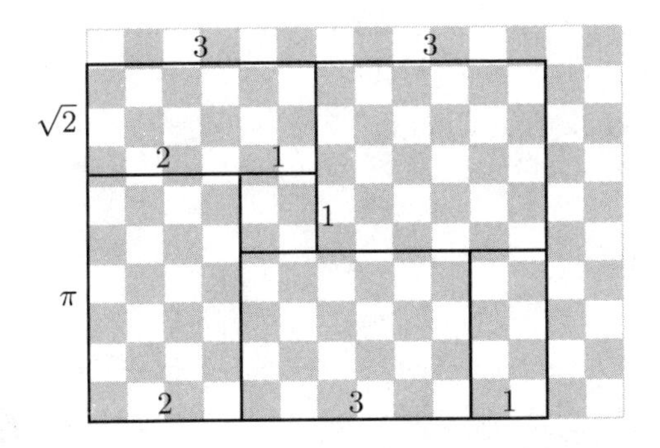

A bit of consideration shows that any rectangle with an integer side contains as much white as black. The big rectangle thus contains as much black as white. Let us assume for a contradiction that none of its sides are integers. It thus conforms to the following type:

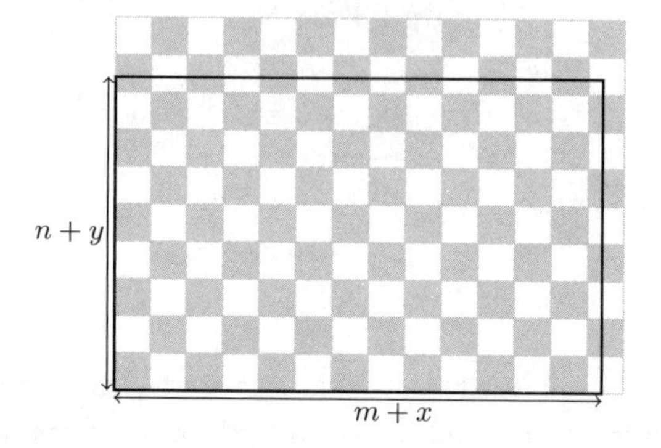


with $m$ and $n$ two integers and $0 < x < 1$, $0 < y < 1$. By removing equal amounts of black and white, we see that the smaller rectangle below must contain equal amounts of black and white:

It is easily shown that this is impossible; hence we have a contradiction. Readers familiar with Fubini's theorem will doubtless appreciate the following method, whose advantage is that it easily generalizes to any arbitrary dimension. We will consider the large rectangle to be $[0, a] \times [0, b]$ in $\mathbb{R}^2$, and the goal becomes to show that either $a$ or $b$ is an integer, given that we can partition $[0, a] \times [0, b]$ into $n$ rectangles $P_1, ..., P_n$, each with at least one integer side.

Consider the following function $F$:

$$F : \begin{cases} \mathbb{R} \times \mathbb{R} & \to \mathbb{C} \\ (x, y) & \mapsto e^{2i\pi(x+y)} \end{cases}$$

For $\alpha, \beta, \gamma, \delta \in \mathbb{R}$ such that $\alpha < \beta$ and $\gamma < \delta$, by applying Fubini's theorem ($F$ is continuous and we integrate over a rectangle), we obtain

$$\iint_{[\alpha,\beta]\times[\gamma,\delta]} F(x,y)\,\mathrm{d}x\,\mathrm{d}y = \left(\int_\alpha^\beta e^{2i\pi x}\,\mathrm{d}x\right)\left(\int_\gamma^\delta e^{2i\pi y}\,\mathrm{d}y\right)$$

$$\iint_{[\alpha,\beta]\times[\gamma,\delta]} F(x,y)\,\mathrm{d}x\,\mathrm{d}y = \frac{e^{2i\pi\beta} - e^{2i\pi\alpha}}{2i\pi} \times \frac{e^{2i\pi\delta} - e^{2i\pi\gamma}}{2i\pi}$$

This integral is 0 if and only if $\beta - \alpha$ or $\delta - \gamma$ is an integer. The integral of $F$ over rectangles $P_1, \ldots, P_n$ is thus 0 and, thus, by linearity, the integral of $F$ over $[0, a] \times [0, b]$ is 0. Thus, $a$ and $b$ are integers.

**90** The electrician, who is in the first house, starts by connecting points 2 and 3 to each other; points 4, 5, and 6 together; and points 7, 8, 9, and 10 together. He then goes to the second house. He wires the battery to point A and the lamp to point B, then to point C, and so on through point J. This allows him to identify some groups. If, for example, the lamp doesn't turn on, it means point A corresponds to point 1. If the lamp only turns on at point C, it means that points A/C correspond to points 2/3, and so on. By doing all the possible tests, that is, by placing the battery on all the possible points and using the lamp to test all others each time, he can identify to which groups of points the groups 1, 2/3, 4/5/6, 7/8/9/10 correspond. Up to renaming the points, we can assume for simplicity that the corresponding groups are A, B/C, D/E/F, G/H/I/J.

Before leaving, he connects the points A/B/D/G together, and points C/E/H together, and points F/I together; he leaves J alone. He then returns to the first house, does the same tests, and finds all the correspondences again. If, for example, points 1/2/6/8 correspond to points A/B/D/G, then he concludes that point 1 corresponds to point A (which he already knew), that point 2 corresponds to point B, that point 6 corresponds to point D, and that point 8 corresponds to point G.

**91** If we split the class into two groups, within each group there will be a given number of in-group knowledge relations. Let us consider the total number of knowledge relations, that is, the number of knowledge relations in the first group plus the number of relations in the second.

If a student has (strictly) more friends in their group than in the other, we change their group. This has as an effect to strictly diminish the total number of knowledge relations. Since this number cannot diminish forever, there is a moment at which no student has more friends in their own group than in the other.

Note that this same proof can be presented nonconstructively. Since there is a finite number of arrangements of the students into two groups, there is an arrangement in which the total relation number is minimal. This arrangement is clearly satisfactory.

**92** After some failed attempts, one can conjecture that there is no strategy that wins more than half the time on average. We will prove this conjecture by two different methods.

**Method 1** A strategy consists of calling red on the basis of the cards seen previously. We can thus consider convoluted strategies, such as "call red on the third card if the first two cards are 2♡, 3♢; otherwise, call red on the fourth card." How should we account for all possible strategies? A simple way is to imagine a gigantic table, with 52! rows and 52 columns, where rows represent all the possible configurations of the deck of cards.

When we mix the deck of cards, we're sure to find one of the lines of the table, and every line can be obtained by a shuffle. Suppose we have a strategy. With this table, we can make it explicit by drawing, on each line of the table, a mark in the column in which red will be called because of the previously revealed cards.

$10\diamondsuit$ $J\spadesuit\,|\,3\clubsuit$ $\cdots\cdots\cdots\cdots\cdots\cdots\cdots\cdots$ $5\heartsuit$ $2\spadesuit$ $1\clubsuit$

$8\spadesuit$ $1\diamondsuit$ $10\clubsuit$ $\cdots\cdot$ $K\clubsuit\,|\,Q\spadesuit$ $\cdots\cdots\cdots\cdot$ $8\clubsuit$ $9\heartsuit$ $9\diamondsuit$

$\vdots$ $\vdots$

$7\heartsuit$ $4\spadesuit$ $9\heartsuit$ $\cdots\cdots\cdots$ $10\heartsuit\,|\,3\spadesuit$ $\cdots\cdot$ $2\clubsuit$ $8\diamondsuit$ $K\diamondsuit$

Note that if a line of the table has a mark, such as after the first two cards ($10\diamondsuit$ and $J\spadesuit$, like in the example above), all the other lines of the table starting with the same two cards must have a mark in the same place. Bob wins if and only if the card after the mark is red. Since the 52! possible configurations are equally likely, the probability of winning is the number of lines where the card after the mark is red, divided by the total number of lines (52!). The problem is therefore to show that there are as many lines where the mark precedes a red card as lines where it precedes a black card. For this, let us modify the table by swapping the last card of each line with the card after the tick. Thus, the configuration

$7\heartsuit$ $4\spadesuit$ $9\heartsuit$ $\cdots\cdots\cdots$ $10\heartsuit\,|\,3\spadesuit$ $\cdots\cdot$ $2\clubsuit$ $8\diamondsuit$ $K\diamondsuit$

becomes

$7\heartsuit$ $4\spadesuit$ $9\heartsuit$ $\cdots\cdots\cdots$ $10\heartsuit\,|\,K\diamondsuit$ $\cdots\cdot$ $2\clubsuit$ $8\diamondsuit$ $3\spadesuit$

One could think we have profoundly changed the table, but in fact we have only reordered the lines! Indeed, the transformation we applied on the lines is a bijection since it is injective (if two lines yield the same new line, they must have been the same before) and surjective (for every line, there is a line which is transformed into this line). Thus, the total number of lines where the mark precedes a red card is equal to the number of new lines where the mark precedes a black card, which is the number of lines where the last card was red, which is $\frac{52!}{2}$. Hence we have proven the result.

**Method 2** Let us show by induction that all strategies are equivalent: we always have a 50/50 chance, regardless of what we do. In a more general argument, we will show that for a deck of $N$ cards containing $N_r$ red cards, all strategies are equivalent and they all yield the same probability $p_N = \frac{N_r}{N}$. In particular, for a standard card deck, we recover the conjectured $p = \frac{1}{2}$.

The induction is done on the total number of cards of the deck. Initialization is obvious for decks of only one card. For the induction, we consider an integer $N \geq 1$ and assume that for any deck of $N$ cards containing $N_r$ red cards (with $N_r$ arbitrary in $\{0, ..., N\}$), all strategies are equivalent and yield the same probability $p_N = \frac{N_r}{N}$. We now consider a pack of $N+1$ cards with $N_r$ red cards and an arbitrary strategy. There are two possibilities: either red is called immediately, and in this case the probability of guessing right is indeed $p_{N+1} = \frac{N_r}{N+1}$, or it lets the first card be revealed.

In the latter case, we have a deck of $N$ cards; if the card that was revealed is red (which occurs with probability $\frac{N_r}{N+1}$), we obtain a probability of $\frac{N_r-1}{N}$. However, if the revealed card is black (which occurs with probability $\frac{N-N_r+1}{N+1}$), we obtain a probability of $\frac{N_r}{N}$. In the end, the probability we sought is

$$p_{N+1} = \left(\frac{N_r}{N+1}\right) \times \left(\frac{N_r - 1}{N}\right) + \left(\frac{N - N_r + 1}{N+1}\right) \times \left(\frac{N_r}{N}\right) = \frac{N_r}{N+1}$$

So, the induction in complete.

**93** First off, it isn't obvious how to do less than $2\sqrt{2}$ (dropping units for legibility). This is possible by digging along two sides, plus half the diagonal, as below:

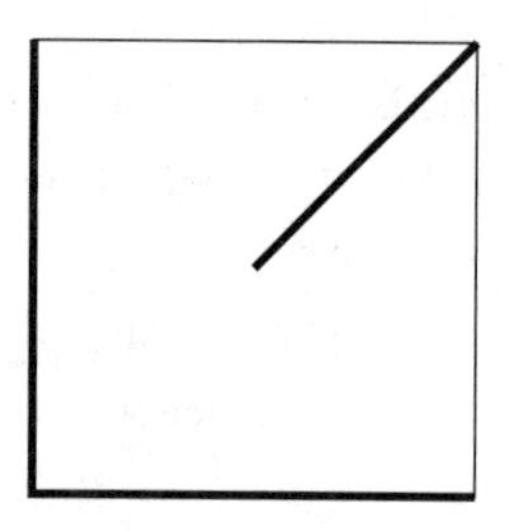

It is easily seen that one is sure to find the cable. The length of the trenches is $\left(2 + \frac{\sqrt{2}}{2}\right) \simeq 2.71$, which is better but still doesn't answer the question. We will modify the configuration slightly as follows:

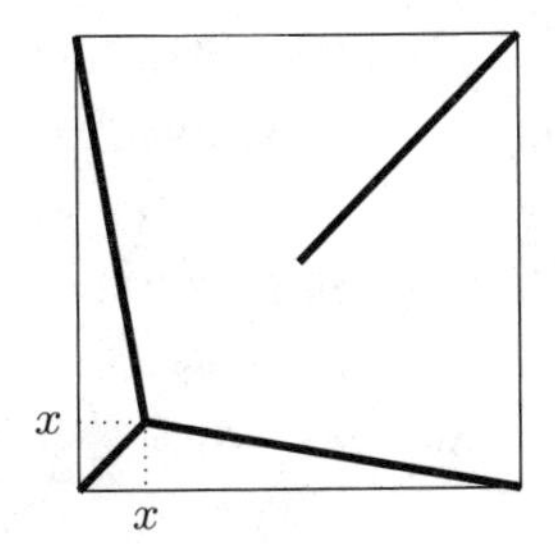


Now, $x$ is between 0 and 0.5, and the goal is to optimize it so that the trench length is minimal. The total length of trenches is

$$f(x) = \frac{\sqrt{2}}{2} + 2\sqrt{(1-x)^2 + x^2} + x\sqrt{2}$$

To find the minimum of $f$ we may differentiate it:

$$f'(x) = 2\frac{2x-1}{\sqrt{(1-x)^2 + x^2}} + \sqrt{2}$$

This derivative is 0 when

$$4x - 2 + \sqrt{2}\sqrt{(1-x)^2 + x^2} = 0$$

which implies

$$x = \frac{3 \pm \sqrt{3}}{6}$$

Without even proving that $f$ in fact attains its minimum at $\frac{3-\sqrt{3}}{6}$, we can simply test the obtained configuration for $x = \frac{3-\sqrt{3}}{6}$, which gives the total length of trench

$$f\left(\frac{3-\sqrt{3}}{6}\right) = \frac{\sqrt{3}+2}{\sqrt{2}} \simeq 2.639 < 2.64$$

and solves the problem.

One might wonder if the configuration obtained this way is optimal. Intuitively, it seems pretty clear; nonetheless, it is to this day an open problem.

It was proposed by Polish mathematician Stefan Mazurkiewicz (1888 – 1945) in the context not of a square, but an arbitrary polygon.[4]

Since then, the only lower bound found is 2, even though everyone is convinced that the solution is $\frac{\sqrt{3}+2}{\sqrt{2}}$. In 2013, this lower bound was improved[5] from 2 to $2+10^{-12}$, and in 2019, it was again improved[6] from $2+10^{-12}$ to 2.0002. Progress has been slow, which shows how hard the problem is.

**94** The earnings cover at most 11 months. Let us show first that the earnings cannot cover more than 12 months. We use contradiction and denote $c_1, c_2, \ldots, c_{12}$ as the earnings of months $1, 2, \ldots, 12$. Consider the sum of the following values:

$$\begin{array}{cccccccc} c_1 & c_2 & c_3 & c_4 & c_5 & c_6 & c_7 & c_8 \\ c_2 & c_3 & c_4 & c_5 & c_6 & c_7 & c_8 & c_9 \\ c_3 & c_4 & c_5 & c_6 & c_7 & c_8 & c_9 & c_{10} \\ c_4 & c_5 & c_6 & c_7 & c_8 & c_9 & c_{10} & c_{11} \\ c_5 & c_6 & c_7 & c_8 & c_9 & c_{10} & c_{11} & c_{12} \end{array}$$

The sum of each row represents the profits over 8 consecutive months, which is thus strictly positive. The sum of each column represents the profits over 5 consecutive months and is thus strictly negative. Thus, the sum of all the entries in the table is strictly positive when all the rows are summed together; however, the sum is strictly negative when all the columns are summed together. This is a contradiction; it is thus impossible that the earnings cover more than 11 months.

The earnings can cover 11 months if we consider the following gain-loss sequence:

$$5, -8, 5, 5, -8, 5, -8, 5, 5, -8, 5$$

[4] See the Wikipedia article https://en.wikipedia.org/wiki/Opaque_set for more information.

[5] A. Dumitrescu and M. Jiang, "The opaque square", *Proceedings of the Thirtieth Annual Symposium on Computational Geometry* (Association for Computing Machinery, 2014): 529–538.

[6] A. Kawamura, S. Moriyama, Y. Otachi, and J. Pach, "A lower bound on opaque sets", *Computational Geometry*, 80 (July 2019): 13–22.

In the general case, with $p$, $q$ coprime, the earnings can cover at most $p+q-2$ months. We show, just as in the case $p = 5$ and $q = 8$, that the earnings cannot cover more than $p + q - 2$ months. Indeed, we can reason by contradiction and write the profits in a table as we did before:

$$\begin{array}{cccc} c_1 & c_2 & \cdots & c_q \\ c_2 & c_3 & \cdots & c_{q+1} \\ \vdots & \vdots & & \vdots \\ c_p & c_{p+1} & \cdots & c_{p+q-1} \end{array}$$

Summing along the rows, then along columns, we obtain the same contradiction. It remains for us to show that the earnings can cover $p + q - 2$ months.

Suppose we have $c_1, c_2, ..., c_{p+q-2}$; let us find a necessary and sufficient condition on the $c_i$ so that the condition on the profits is satisfied. Let $s_0 = 0$ and, for $1 \le k \le p+q-2$, let

$$s_k = \sum_{i=1}^{k} c_i$$

Let us assume the condition on the profits is satisfied. For $0 \le k \le p-2$, we have $s_{k+q} - s_k > 0$ since it is a sum of $q$ consecutive profit. Likewise, for $0 \le k \le q-2$, $s_{k+p} - s_k < 0$. It is clear that this condition on the $s_k$ is sufficient. Let us notice that constructing the $s_k$ is sufficient since we will always be able to find the corresponding $c_k$ by taking $c_k = s_k - s_{k-1}$.

Let us see how this allows us to exhibit an example in the case in which $p = 5$ and $q = 8$. We necessarily have that $s_7 < s_2 < s_{10} < s_5 < s_0 < s_8 < s_3 < s_{11} < s_6 < s_1 < s_9 < s_4$. To satisfy these conditions, we can assign to $s_k$ the following values:

| $s_7$ | $s_2$ | $s_{10}$ | $s_5$ | $s_0$ | $s_8$ | $s_3$ | $s_{11}$ | $s_6$ | $s_1$ | $s_9$ | $s_4$ |
|---|---|---|---|---|---|---|---|---|---|---|---|
| $-4$ | $-3$ | $-2$ | $-1$ | 0 | 1 | 2 | 3 | 4 | 5 | 6 | 7 |

Thus, the $s_k$ satisfy the above inequalities. Moreover, all inequalities of the type $s_k < s_{k+q}$ and $s_{k+p} < s_k$ have been written. Thus, the $c_k$ values associated with these $s_k$ values satisfy the profit conditions. In fact we have created the same example as before: $5, -8, 5, 5, -8, 5, -8, 5, 5, -8, 5$.

Let us now try to adapt this method to the general case. We will assume without loss of generality that $p < q$. As before, let us show that the $p+q-2$ inequalities of type $s_k < s_{k+q}$ or $s_{k+p} < s_k$ imply some ordering of the $s_k$ and, conversely, that this ordering implies the inequalities. This is sufficient to draw a conclusion, although it will remain for us to assign to each $s_k$ the integer representing its rank in the ordering. Thus, the $s_k$ values will satisfy the $p+q-2$ inequalities and their associated $c_k$ values will satisfy the profit condition.

**Implication** We assume that the $p+q-2$ inequalities of type $s_k < s_{k+q}$ or $s_{k+p} < s_k$ are satisfied; the goal is to order the $s_k$. First, the smallest of the $s_k$ has to be $s_{q-1}$. Indeed, for any integer $m$ in $\{0, \ldots, p+q-2\}$ such that $m \neq q-1$, there is $m'$ such that $m' = m-q$ or $m' = m+p$ and thus $s_{m'} < s_m$. Let us thus denote $u_0 = q-1$ ($u_k$ will be the index such that $s_{u_k}$ will be the $k$th smallest element of the family of the $s_i$). For $k \in \{0, \ldots, p+q-3\}$ such that $u_k$ is well defined and not equal to $p-1$, we define $u_{k+1}$ as the unique integer between 0 and $p+q-2$ such that $u_{k+1} = u_k + q$ or $u_{k+1} = u_k - p$. We thus have $s_{u_k} < s_{u_{k+1}}$. It remains to show that we can construct an ordering of the $s_k$ in this way: that we can construct $u_k$ up to $u_{p+q-2} = p-1$ and that the $u_k$ values thus constructed are all different.

Let $m \leq p+q-2$ such that $u_0, u_1, \ldots, u_m$ have been constructed. Let us show that all these $u_k$ are different. Suppose, for a contradiction, that there are two integers $i < j$ between 0 and $p+q-2$ such that $u_i = u_j$. By construction of the $u_k$, there are two integers $s, t$ such that $u_j = u_i - ps + qt$. Thus, $p$ divides $qt$, but because $p$ is coprime with $q$, $p$ divides $t$; thus, $t \geq p$. Likewise, $q$ divides $s$, and, thus, $s \geq q$. Hence, $s+t \geq q+p$. Moreover, since $s$ and $t$ are not both zero and $ps = qt$, $s$ and $t$ must both be nonzero. But $j - i = s + t$ by construction, so $j - i \geq p + q$. This is a contradiction since $i$ and $j$ are both between 0 and $p+q-2$.

It remains only to show that for $k \leq p+q-3$ such that $i_k$ is constructed, $u_k \neq p-1$. This will ensure we can construct $u_0, u_1, \ldots, u_{p+q-2}$, with the $u_k$ all different. Assume, for a contradiction, that there is $k \leq p+q-3$ such that $u_k = p-1$. We have $s$ and $t$ two integers such that $u_k = u_0 - ps + qt$.

This thus implies $p(s+1) = q(t+1)$. As before, $s+1+t+1 \geq q+p$; hence, $k \geq p+q-2$ is a contradiction. Hence, we have proven our result.

**Converse** Let us assume conversely, with the above notation, that

$$s_{u_0} < \cdots < s_{u_{p+q-2}} .$$

This implies $p+q-2$ inequalities, all different and of type $s_k < s_{k+q}$ or $s_{k+p} < s_k$; thus, it does indeed imply the $p+q-2$ inequalities.

**95** The first idea that might come to mind is to number the squares from 0 to 63 and to compute the sum mod 64 of the squares, counting the number on a square only if the coin on it is tails. In this way, any initial condition of the board encodes a number, and we would claim that to access the number of the secret square, Ivan only needs to turn over the correct coin. For example, if at the start, the board encodes the number 43 and the secret square is number 52, Ivan only needs to turn over the coin located on number 9.

Unfortunately, this method fails if the coin on square number 9 is already on tails. Flipping it will mean subtracting 9 instead of adding 9 as wanted. Ivan could also have flipped coin 55 from tails to heads since $-55 \equiv 9\,[64]$, but if this coin is already on heads, he is stuck.

To avoid this issue, it suffices to number squares 0 to 63 in base 2 (see reminder 7):

$$000000_2, 000001_2, 000010_2, \ldots, 111111_2$$

Then, define the following addition (known as XOR): we follow the normal digit-wise addition rules, except that $1+1=0$ and we do not carry units over to the next digit. For example

$$101100_2 + 111111_2 = 010011_2$$

and

$$001101_2 + 011001_2 = 010100_2$$

Thus, adding or subtracting the same number with this rule is the same, and the problem has disappeared.

Ivan computes first the number of the square encoded by the chess board by summing all the squares. Then, he adds what he must add to reach the number of the secret square (which is the digits that need to change). For example, if the sum is $100110_2$ and the secret square is $011100_2$, one must add $111010_2$. Whether this coin goes from heads to tails or vice versa will have the same effect. Dimitri then computes the number of the square encoded by the modified chessboard and he finds $011100_2$, the number of the secret square.

Let us now see that there is no winning strategy if the number of squares is not a power of two. We consider the game board with $n$ squares. Let us place ourselves directly in the shoes of Dimitri, who sees the chessboard for the first time and must guess the secret square. He must thus have established with Ivan a code in which each combination of coins corresponds to a square, thus a number between 1 and $n$. Denote $c_i$ as the number of combinations that encode the square $i$ for $i$ running from 1 to $n$. There are $2^n$ possible combinations of coins, so $\sum_{k=1}^{n} c_i = 2^n$.

Moreover, and this is the key point of the code, one must be able to access a combination coding $i$ starting from any combination. Looking at things the other way, every combination must be reachable from one of the $c_i$ combinations encoding $i$. Since each combination encoding $i$ generates at most $n$ different combinations, we must have $n \times c_i \geq 2^n$. Thus,

$$\sum_{i=1}^{n} c_i = 2^n \quad \text{and} \quad \forall i \in \{1, ..., n\},\ c_i \geq \frac{2^n}{n}$$

so,

$$\forall i \in \{1, ..., n\}, c_i = \frac{2^n}{n}$$

However, the $c_i$ are integers; so $n$ divides $2^n$ and $n$ must be a power of 2.

**96** When the principal chooses a sequence $u$ of real numbers (whose $n$th term corresponds to the real number of the $n$th student), the students must announce a sequence $v$ (whose $n$th term corresponds to the real number announced by the $n$th student), such that $v$ is equal component-wise to $u$ except for finitely many terms. This leads to considering the equivalence relation (see reminder 3), which states that two sequences are equivalent if they are equal after some rank:

$$\forall (u, v) \in \left(\{0, 1\}^{\mathbb{N}^*}\right)^2, u \sim v \Longleftrightarrow \exists N \in \mathbb{N}^*, \forall n \geq N, u_n = v_n$$

Students agree to assign to each equivalence class a representative (we use the axiom of choice here). When the principal chooses a sequence $u$, if the students manage to announce the representative $v$ of its equivalence class, since $u \sim v$, the game is over. But how? It suffices for the student of step $n$ to complete the sequence they observe by adding $n$ zeros to the start. In this way, they form an equivalent sequence to $u$ (since aside from the zeros at the start, it is equal to $u$ after rank $n$), which is thus represented by the same sequence $v$. It remains for the student to announce $v_n$.

**97** In riddle 39, we considered the parity of the number of black hats seen by the student at the top. Of course, this method is not possible here since there may be infinitely many black hats. What remains unchanged is that the student at the top must announce a color depending on all the hats she sees, which means depending on the sequence with values in $\{0,1\}$ whose $n$th term ($n \geq 1$) corresponds to the color of the hat of student number $n$.

A strategy thus consists in fixing a partition of $\{0,1\}^{\mathbb{N}^*}$ into two subsets $A$ and $B$. Set $A$ contains all sequences for which the top student announces "white," set $B$ corresponds to those for which she announces "black." Under what condition does this partition constitute a working strategy? We note that it is necessary that for any sequence in set $A$, if we change exactly one term, we end up in set $B$; we will note this as condition $\mathcal{P}$. Indeed, if two sequences of $A$ are identical up to the $n$th term, the student on step $n$ will have no way to guess his color. The condition $\mathcal{P}$ is sufficient since, if a partition satisfying $\mathcal{P}$ exists, then the students can deduce in turn their color just as in riddle 39.

Let us now construct a partition that satisfies $\mathcal{P}$. If the zero sequence is in, say, set $A$, we already know that all the sequences that are obtained by changing an even number of terms in the zero sequence are in $A$ too. Unfortunately, these are all zero sequences after some rank: by starting with the zero sequence, we can't seem to define in which set the constant sequence with all terms equal to one should lie, for example. We are thus led to consider the same equivalence relation as in the previous riddle, that is, two sequences are equivalent if they are equal after some rank:

$$\forall (u, v) \in \left(\{0, 1\}^{\mathbb{N}^*}\right)^2, u \sim v \Longleftrightarrow \exists N \in \mathbb{N}^*, \forall n \geq N, u_n = v_n$$

It is easy to check that this is indeed an equivalence relation. In each equivalence class, let us take a representative (using the axiom of choice once more), and let's say it is in $A$. Now, let us take an arbitrary sequence $u \in \{0, 1\}^{\mathbb{N}^*}$. It is in an equivalence class that has a representative $v$: $u \sim v$. The sequences $u$ and $v$ must thus differ in only a finite number of terms. Either this number is even and we fix $u \in A$, or it is odd and we fix $u \in B$.

We have thus constructed a partition of $\{0, 1\}^{\mathbb{N}^*}$ satisfying $\mathcal{P}$. The strategy consists of the students agreeing on an encoding of the type "white" $= 0$ and "black" $= 1$, on the construction of sets $A$ and $B$, and finally on encoding "white" $= A$ and "black" $= B$ for the first student. Thus, the students can deduce in turn the color of their hats. For example, if the students know that the zero sequence is in $A$, that the first student announced "white" and that the second student sees in front of him only white hats, he knows his hat is white, because otherwise the first student would have seen the sequence (1,000...), which is in $B$ and would have announced "black." The second student thus announces his own color, and the followings students will do the same by following the same reasoning as in riddle 39.

**98** With riddle 96, we saw that the equivalence relation $\sim$ allows the students, very surprisingly, to guess the numbers on their own heads. The problem with this method is that we don't know from which rank the sequence and its representative are equal. The trick is to partition the drawers into 100 disjoint sequences, such as the sequences $u^m = (100n+m)_{n\in\mathbb{N}}$, where $0 \leq m \leq 99$. From now on, we denote $\widetilde{n}$ the real number of drawer $n$. For $m$ fixed, by opening the sequence of drawers $u^m$ (meaning drawers numbered $u_n^m$, where $n \in \mathbb{N}$), we discover the sequence of real numbers $\left(\widetilde{u_n^m}\right)_{n\in\mathbb{N}}$. This sequence is equal component-wise to its representative after some rank $r_m$. Students may agree on the following strategy: student number $m$ looks at all sequences $u^k$ except $u^m$. They can thus compute $r = \max\left(r_k\right)_{k\neq m}$. Then that student opens the sequence $u^m$ from rank $r + A$. Thus, they know almost the whole sequence $\left(\widetilde{u_n^m}\right)_{n\in\mathbb{N}}$: only the $r + 1$ first terms are unknown to them. Nevertheless, they can find its representative by following the solution to riddle 96, which is to complete the observed sequence with zeros at the start. Since $\underbrace{0,\ldots,0}_{r+1 \text{ zeros}}, \widetilde{u^m_{r+1}}, \widetilde{u^m_{r+2}}, \widetilde{u^m_{r+3}}, \ldots$ is equivalent to $\left(\widetilde{u^m_n}\right)_{n\in\mathbb{N}}$, it has the same representative $v$. The student announces $v_r$ for drawer $u^m_r$. For their prediction to be correct, it suffices that $r_m \leq r$. This is always the case, unless $r_m$ is a strict maximum of $\{r_k\}_{0\leq k\leq 99}$. With this set having at most one strict maximum, there is indeed at most one student who is mistaken.

**99** Let $\mathcal{P}$ be a finite set of points in the 2D plane satisfying the following conditions: for all $A, B \in \mathcal{P}$ such that $A \neq B$, there is $C \in \mathcal{P}$ such that $C \neq A$, $C \neq B$, and $C \in (AB)$. One needs to show that all points of $\mathcal{P}$ form a line.

For a contradiction, let us suppose that they do not all form a line. Let $M \in \mathcal{P}$, so that there is at least one line $\Delta$ passing through two points of $\mathcal{P}$ such that $M \notin \Delta$ (otherwise all points would form a line). Let us denote $\mathcal{D}_M$ as the set of lines formed by at least two points of $\mathcal{P}$ that don't include $M$. This set is nonempty and clearly finite. We can thus define

$$m(M) = \min\left\{d\left(M, \Delta\right) | \Delta \in \mathcal{D}_M\right\}$$

with $d\left(M, \Delta\right)$ as the distance from point $M$ to the line $\Delta$, that is, the distance $MH$ if $H$ denotes the orthogonal projection of $M$ on $\Delta$.

Consider now the point $M$ of $\mathcal{P}$ such that $m(M)$ is minimal. Recycling above notation, we note $\Delta$ as the line of $\mathcal{D}_M$ such that $m(M) = d\left(M, \Delta\right)$. Let $H$ be the orthogonal projection of $M$ onto $\Delta$. There are two points $A, B \in \mathcal{P}$ such that $\Delta = (AB)$, and we have a third point $C$ distinct from $A$ and $B$ belonging to $\Delta$.

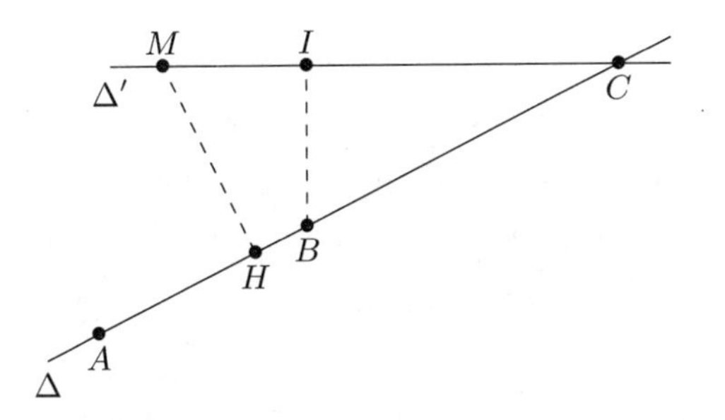

At least two points are on the same side of $H$; here, these are the points $B$ and $C$. If $I$ is the orthogonal projection of $B$ on $(MC)$, it is clear that $BI < MH$, which contradicts the minimality of $m(M)$. Hence, we have proven the result.

**100** We will begin by showing that we can reach the fourth line. Then we will show that we can't reach the fifth line, a proof that will be much more delicate.

The configurations below allow one to reach the second, third, and fourth lines, respectively.

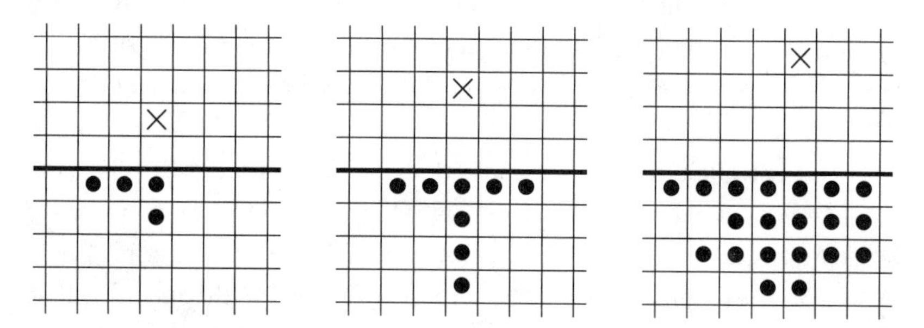

To reach the third line, we build on the configuration used to reach the second line. The steps are shown below:

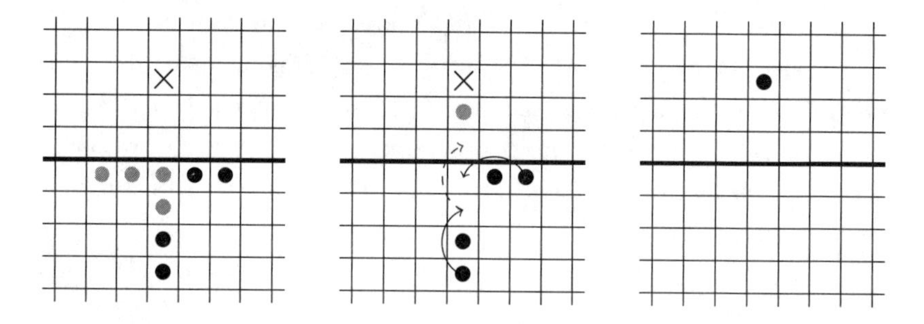

Reaching the fourth line turns out to be a bit more complicated. We put down the configuration necessary to reach the third line, which already gives us a marble on the third line after some moves. The other marbles allow us to place one on the second line, which finally allows us to move it to the fourth line:

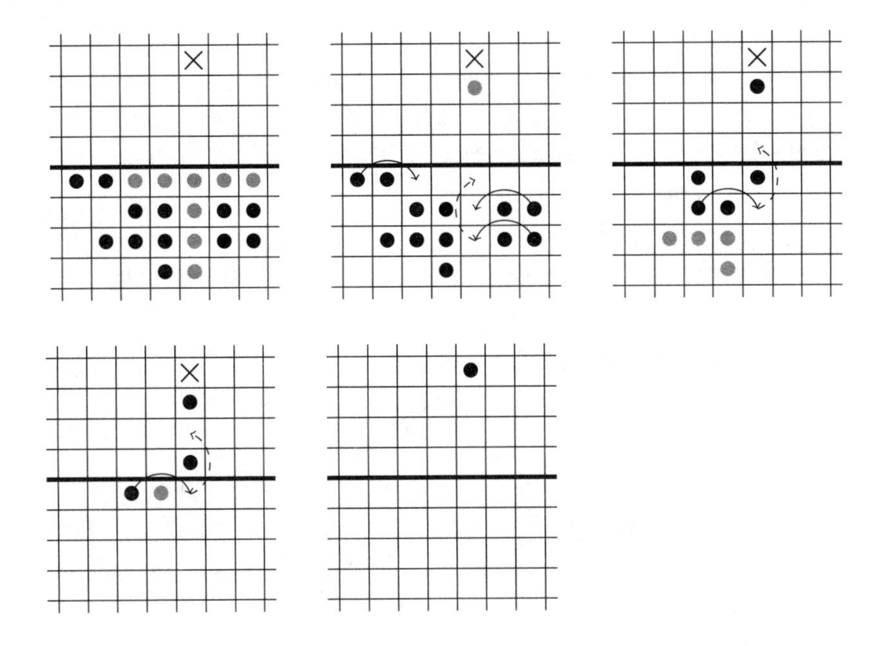

It remains for us to show that we cannot reach the fifth line. The following proof is owed to the famous British mathematician, the late John Horton Conway.

Let us begin by assigning the value 1 to a square on the fifth line. We assign to other squares the value $\phi^r$ where $r$ is the shortest number of steps (vertical or horizontal) to reach the square of value 1 and $\phi$ is the golden ratio minus 1: $\phi \simeq 0.618$. More accurately, $\phi$ is the only positive real number satisfying $\phi^2 + \phi = 1$. A graphical representation is as follows:

| $\phi^3$ | $\phi^2$ | $\phi^1$ | 1 | $\phi^1$ | $\phi^2$ | $\phi^3$ |
|---|---|---|---|---|---|---|
| $\phi^4$ | $\phi^3$ | $\phi^2$ | $\phi^1$ | $\phi^2$ | $\phi^3$ | $\phi^4$ |
| $\phi^5$ | $\phi^4$ | $\phi^3$ | $\phi^2$ | $\phi^3$ | $\phi^4$ | $\phi^5$ |
| $\phi^6$ | $\phi^5$ | $\phi^4$ | $\phi^3$ | $\phi^4$ | $\phi^5$ | $\phi^6$ |
| $\phi^7$ | $\phi^6$ | $\phi^5$ | $\phi^4$ | $\phi^5$ | $\phi^6$ | $\phi^7$ |
| $\phi^8$ | $\phi^7$ | $\phi^6$ | $\phi^5$ | $\phi^6$ | $\phi^7$ | $\phi^8$ |
| $\phi^9$ | $\phi^8$ | $\phi^7$ | $\phi^6$ | $\phi^7$ | $\phi^8$ | $\phi^9$ |

Let us assume for contradiction that it is possible to reach square 1. We thus have a configuration containing a finite number of marbles below the line, which after some number of steps leads to a marble on square 1. We call the energy of a marble the value of the square on which it lies, and total energy is the sum of the energies of marbles.

Let us show that the total energy cannot increase when a move is made. There are three kinds of moves:

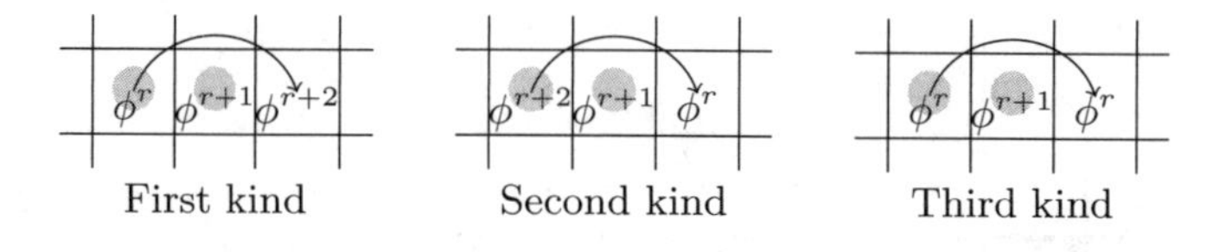


- For the first kind of move, the energy change is $\phi^{r+2} - \left(\phi^{r+1} + \phi^r\right)$. Since $0 < \phi < 1$, $\phi^{r+2} < \phi^{r+1} < \phi^{r+1} + \phi^r$. The energy strictly decreases.

- For the second type, the energy change is $\phi^r - \left(\phi^{r+1} + \phi^{r+2}\right)$. Since $\phi^2 + \phi = 1$, $\phi^{r+2} + \phi^{r+1} = \phi^r$, the total energy doesn't change.

- Finally, for the third type, the variation is $\phi^{r+1} - \left(\phi^r + \phi^{r+1}\right) = -\phi^r < 0$, so the energy strictly decreases.

In the end, whatever one does, the energy is nonincreasing. Let us now compute the energy $E$ contained by the whole lower half of the plane.

The energy of the first line is

$$\phi^5 + 2\phi^6 + 2\phi^7 + 2\phi^8 + ... = \phi^4 \left(\phi + 2\left(\phi^2 + \phi^3 + \phi^4 ...\right)\right)$$

But,

$$\phi^2 + \phi^3 + \phi^4 + ... = \frac{\phi^2}{1 - \phi} = 1$$

and thus the energy of the first line is

$$\phi^4 (\phi + 2)$$

This means the total energy is

$$E = \phi^4 (\phi + 2) \left(1 + \phi + \phi^2 + ...\right)$$

but,

$$1 + \phi + \phi^2 + \phi^3 + ... = \frac{1}{1 - \phi} = \frac{1}{\phi^2}$$

and thus

$$E = \phi^4 (\phi + 2) \times \frac{1}{\phi^2} = \phi^2 (\phi + 2) = (1 - \phi) (\phi + 2) = 2 - \phi - \phi^2 = 1$$

The total energy (of marbles, that is) is strictly less than $E$ since we assumed there was only a finite number of marbles. However, once a marble is on square 1, the total energy of marbles is greater than or equal to 1. This is absurd since energy cannot increase. Thus we conclude that it is impossible to reach the fifth line.

# Part IV

# Reminders

# 1 The Pigeonhole Principle

## Statement of the Principle

The pigeonhole principle is simple and very useful. Let us begin with a statement and an example:

> If 11 socks are put away in 10 drawers, then there is at least one drawer containing (at least) two socks.

It is common sense, and anyone can easily convince themselves of the truth of this principle. One can replace the numbers 10 and 11 chosen in this example by $n$ and $n+1$ where $n$ is any nonzero integer. We obtain the following statement:

> Let $n$ be a positive integer. If $n+1$ socks are put away in $n$ drawers then there is at least one drawer containing at least two socks.

## Examples of Applications

Let us consider a lecture hall which contains 400 students. Let us show there are at least two students who are born on the same day of the year. We apply the pigeonhole principle to the 365 days of the year. (We don't consider leap years.) Each day of the year corresponds to a drawer, and each student, to a sock. Since there are more socks than drawers, there is at least one drawer with two socks. Therefore, there is at least one day of the year where two birthdays will be held. We have the same results with 366 students. But we cannot draw this conclusion if there are 365 students or fewer in the lecture hall.

Here is a more mathematical and more delicate example. Let $x$ be any real number and $N$ a nonzero integer. We want to show that there are two integers $p$ and $q$ with $q \neq 0$ such that $|qx - p| < \frac{1}{N}$.

We recall that the floor of a real number is the largest whole number that is smaller than the real number being considered and is denoted as $\lfloor\ \rfloor$. For example, $\lfloor 4.6 \rfloor = 4$ since 4 is the largest whole number smaller than 4.6, $\lfloor 19 \rfloor = 19$ and $\lfloor -3.2 \rfloor = -4$.

The fractional part of a real number $x$ is $x - \lfloor x \rfloor$, and is denoted as $\{x\}$. For example $\{4.6\} = 4.6 - 4 = 0.6$, $\{19\} = 0$, and $\{-3.2\} = -3.2 + 4 = 0.8$. Let us notice that the fractional part is a nonnegative number and is strictly less than 1. Let us now proceed with the proof of the announced result.

Let us consider the $N + 1$ numbers:

$$\{0 \times x\},\ \{1 \times x\},\ ...,\ \{N \times x\}$$

which all belong to the interval $[0, 1)$. We place them into the $N$ drawers:

$$\left[0, \frac{1}{N}\right),\ \left[\frac{1}{N}, \frac{2}{N}\right),\ ...,\ \left[\frac{N-1}{N}, 1\right)$$

There is therefore a drawer in which there are at least two of these numbers. Hence, there exists a whole number $k$ between 0 and $N - 1$ and two whole numbers $n \neq m$ between 0 and $N$ such that $\{nx\}$ and $\{mx\}$ are in the same drawer $\left[\frac{k}{N}, \frac{k+1}{N}\right)$:

$$\frac{k}{N} \leq \{nx\} < \frac{k+1}{N} \quad \text{and} \quad \frac{k}{N} \leq \{mx\} < \frac{k+1}{N}$$

We deduce that:

$$|\{nx\} - \{mx\}| < \frac{1}{N}$$

Meanwhile, $\{nx\} = nx - \lfloor nx \rfloor$ and $\{mx\} = mx - \lfloor mx \rfloor$; thus

$$\{nx\} - \{mx\} = (n - m)x - (\lfloor nx \rfloor - \lfloor mx \rfloor)$$

Denoting $p$ as the whole number $\lfloor nx \rfloor - \lfloor mx \rfloor$ and $q$ as the nonzero whole number $n - m$ we obtain:

$$|qx - p| < \frac{1}{N}$$

The theorem thus proven is due to Dirichlet, a 19th century German mathematician, who formalized the pigeonhole principle. Therefore, this principle is sometimes known as “Dirichlet’s pigeonhole principle.”

# Comparison of Outcomes to Hypotheses

The method of comparing outcomes to hypotheses is an application of the pigeonhole principle that is used to solve many riddles. To explain what it is, let us take a simple example.

Consider the following problem: nine weights, numbered 1 to 9, that all have the same mass except one, which is heavier. The goal is to find which weight is heavier than the others. To do so, we have a set of Roberval scales with which we can only do two weighings. How do we fulfill the goal under this constraint?

It is easy! Weigh weights 1, 2, and 3 on the right side and weights 7, 8, and 9 on the left side. This weighing allows us to determine which group of three weights contains the heavier weight. We then weigh two weights of this group against one another and leave the third aside. We thus determined which was the heavier weight.

Let us now study a more complicated problem that will use the method of comparing outcomes to hypotheses. We have 10 weights, numbered 1 to 10, that all have the same mass except one, which is heavier. The goal is still to find which is the heavier weight using the Roberval scale only twice.

It is impossible! This might seem surprising, since this problem only has one more weight, but we will use the method of comparing outcomes to hypotheses to show it is impossible. As the name suggests, one must compare the number of hypotheses—possible answers to the problem—to the number of possible outcomes from the experiment we are performing. The problem has 10 possible answers: the heaviest weight can be number 1, or 2, or 3, ..., all the way to 10. So, we can draw up 10 mutually exclusive hypotheses for the experiment. On the other hand, the experiment admits nine possible outcomes: two consecutive weighings with three possible outcomes (scales tip left, or right, or are stable) each is $3 \times 3 = 9$ total outcomes.

We now apply the pigeonhole principle: there are 10 hypotheses and nine possible outcomes, so there are at least two hypotheses that are consistent with the same outcome and therefore cannot always be separated. In other words, there is always at least one outcome of the experiment that leaves you unable to decide between two hypotheses (i.e., between two weights) after the experiment. Therefore, the problem with 10 weights is impossible.

## Riddles Using the Pigeonhole Principle

Riddle 20 is a direct application of the pigeonhole principle. The method of comparing outcomes to hypotheses, which was presented in the previous paragraph, is often used to show the optimality of a given strategy. Riddles 37, 57, 68, and 76 use this method.

# 2 Binomial Coefficients

## Computation of "$n$ choose $p$"

To understand what binomial coefficients are, let us begin with a simple example: how many different ways are there to choose two pencils among four different pencils? We can number the pencils from 1 to 4, and the two chosen pencils can be

- pencils 1 and 2
- pencils 1 and 3
- pencils 1 and 4
- pencils 2 and 4
- pencils 2 and 3
- pencils 3 and 4

There are thus six ways to choose two pencils among four, which means there are six subsets of two elements of the set of pencils with four elements. We therefore use this denotation: $\binom{4}{2} = 6$.

In general, for an integer $n$ and an integer $p$ less than or equal to $n$, we write $\binom{n}{p}$ and read "$n$ choose $p$" as the number of subsets with $p$ elements of a set containing $n$ elements. $\binom{n}{p}$ is called a binomial coefficient.

Let us compute $\binom{n}{p}$. We imagine we have $n$ objects before us and that we need to choose $p$. We can choose the first object however we want: we have $n$ options. For the second object, we can only choose from $n-1$ objects because one object has already been chosen. For the third, we have $n-2$ options, and so on: there are $n-(p-1)$ options for the $p$th object. Thus, there are $n \times (n-1) \times (n-2) \times \cdots \times (n-p+1)$ configurations, which is not the same as choosing a subset with $p$ elements, because for a configuration, we take into account the order of the elements. For a subset of $p$ given elements, there are $p \times (p-1) \times \cdots \times 1$ possible configurations, because there are $p$ options for the first, $p-1$ for the second, and so on. One must thus divide the number $n \times (n-1) \times (n-2) \times \cdots \times (n-p+1)$ by $p \times (p-1) \times \cdots \times 1$ (which is noted $p!$ and is read as the "factorial of $p$" or as "$p$ factorial" with the convention $0! = 1$) to find the number of subsets with $p$ elements:

$$\binom{n}{p} = \frac{n \times (n-1) \times \cdots \times (n-p+1)}{p!} = \frac{n!}{(n-p)!p!}$$

Of course, this formula allows us to find once again the quantity $\binom{4}{2}$ previously computed:

$$\binom{4}{2} = \frac{4!}{(4-2)! \times 2!} = \frac{4 \times 3 \times 2 \times 1}{2 \times 2} = 6$$

## Some Useful Properties

A first obvious formula is the following:

$$\binom{n}{p} = \binom{n}{n-p}$$

Indeed,

$$\binom{n}{n-p} = \frac{n!}{(n-(n-p))!(n-p)!} = \frac{n!}{p!(n-p)!} = \binom{n}{p}$$

This highlights the fact that choosing a set of $p$ elements is the same as choosing the remaining $n-p$ elements. The following formula is fundamental:

$$\binom{n}{p} + \binom{n}{p+1} = \binom{n+1}{p+1}$$

Let us prove it:

$$\binom{n}{p} + \binom{n}{p+1} = \frac{n!}{(n-p)!}\left(\frac{1}{p!} + \frac{n-p}{(p+1)!}\right)$$

$$\binom{n}{p} + \binom{n}{p+1} = \frac{n!}{(n-p)!} \times \frac{p+1+n-p}{(p+1)!}$$

$$\binom{n}{p} + \binom{n}{p+1} = \frac{(n+1)!}{(n-p)!(p+1)!}$$

This allows for simple calculation of binomial coefficients. We can write them as a triangle with, on the $n$th line, the coefficients $\binom{n}{0}, \ldots, \binom{n}{n}$. This gives:

$$\begin{array}{ccccccccccccccccc}
 & & & & & & & & 1 & & & & & & & & \\
 & & & & & & & 1 & & 1 & & & & & & & \\
 & & & & & & 1 & & 2 & & 1 & & & & & & \\
 & & & & & 1 & & 3 & & 3 & & 1 & & & & & \\
 & & & & 1 & & 4 & & 6 & & 4 & & 1 & & & & \\
 & & & 1 & & 5 & & 10 & & 10 & & 5 & & 1 & & & \\
 & & 1 & & 6 & & 15 & & 20 & & 15 & & 6 & & 1 & & \\
 & 1 & & 7 & & 21 & & 35 & & 35 & & 21 & & 7 & & 1 & \\
1 & & 8 & & 28 & & 56 & & 70 & & 56 & & 28 & & 8 & & 1
\end{array}$$

Each line is obtained easily from the previous one: in view of the previous formula, it suffices to sum two consecutive coefficients of the same line to get the one beneath.

Binomials are useful to expand the expression $(a+b)^n$, where $a$ and $b$ are two real numbers and $n$ is a nonzero integer. For example, for $n = 2$, $(a+b)^2 = a^2 + 2ab + b^2$. This is a remarkable equality that one learns in high school. The binomial theorem is a remarkable equality for $n$ arbitrary. It states:

$$(a+b)^n = \sum_{k=0}^{n} \binom{n}{k} a^k b^{n-k}$$

This is easily shown by induction. We can also notice that $\binom{n}{k}$ is the number of ways of a coefficient $a^k b^{n-k}$ in the expansion of $(a+b)^n$.

# Riddles Using Binomial Coefficients

Binomial coefficients are very useful for all combinatorial questions. So, riddles 36, 67, 72, and 76 use binomial coefficients.

# 3 Equivalence Relations

## Definition of Equivalence Relations

The concept of equivalence relation is used only in some particularly interesting riddles of level 3. It is nonetheless a very common notion in mathematics. The idea is to relate elements of a set to each other and thus to gather them into what we call classes.

Let us begin with a simple example. We consider a set of students in a high school, and we introduce the following relation: "Two students are related if they are born in the same month." This relation allows us to group students into classes. To be an equivalence relation, that is to allow for the grouping of elements of a set into classes, a relation must be reflexive, symmetric, and transitive. The relation of the example is in fact an equivalence relation since it verifies these three conditions:

- Every student is related to themself (since they are born in the same month as themself); this is reflexivity.
- If student $A$ is related to student $B$, then student $B$ is related to student $A$; this is symmetry.
- If student $A$ is related to student $B$ and student $B$ is related to student $C$ then student $A$ is related to student $C$; this is transitivity.

Formally, an equivalence relation on a set $E$ is often denoted as $\sim$. For two students $x$ and $y$ of $E$, we will note $x \sim y$ if $x$ is related to $y$. For a relation to be an equivalence relation, it must verify the three cited properties:

- $\forall x \in E, x \sim x$: $x$ must be related to itself; this is reflexivity.
- $\forall (x, y) \in E^2, x \sim y \Rightarrow y \sim x$: if $x$ is related to $y$, then $y$ is related to $x$; this is symmetry.
- $\forall (x, y, z) \in E^3, (x \sim y, y \sim z) \Rightarrow x \sim z$: if $x$ is related to $y$ and if $y$ is related to $z$, then $x$ is related to $z$; this is transitivity.

## Examples of Equivalence Relations

For $\mathbb{Z}$, the set of whole numbers, we can define the following relation:

$$\forall (x, y) \in \mathbb{Z}^2, x \sim y \text{ if } \exists k \in \mathbb{Z}, x - y = 3k$$

which means $x$ is related to $y$ if $x - y$ is a multiple of 3. This is indeed an equivalence relation since:

- For $x$ in $\mathbb{Z}$, $x$ is related to $x$ since $x - x = 0$ is a multiple of 3: $0 = 3 \times 0$.
- For $x$ and $y$ in $\mathbb{Z}$, if $x - y$ is a multiple of 3 then $y - x$ is also a multiple of 3. (It suffices to change the sign of $k$.) Hence, if $x \sim y$, then $y \sim x$.
- For $x$, $y$, and $z$ in $\mathbb{Z}$, if $x \sim y$ and $y \sim z$ then there is $k_1$ such that $x - y = 3k_1$ and there is $k_2$ such that $y - z = 3k_2$; thus, $x - z = 3(k_1 + k_2)$ and $x \sim z$.

In fact, we simply defined the relation "to be congruent modulo 3"; perhaps review congruence (see reminder 6).

## Equivalence Classes and Representatives

When a set $E$ is endowed with an equivalence relation $(\sim)$, we can regroup its elements into classes. For a given element $x$ of $E$, the class of $x$ noted as $Cl(x)$ is the set of elements $y$ of $E$ that are related to $x$:

$$Cl(x) = \{y \in E | x \sim y\}$$

We note $E/\sim$ the set of classes of elements of $E$:

$$E/\sim = \{Cl(x) | x \in E\}$$

$E/\sim$ is thus a set of sets. For $x, y \in E$ such that $x \sim y$, $Cl(x) = Cl(y)$. Indeed, for $z \in E$, if $z \sim x$, by transitivity, $z \sim y$, and conversely: $z \sim x \Longleftrightarrow z \sim y$.

For example, for the relation "to be congruent modulo 3" (as previously stated), let us determine the equivalence classes.

$Cl(0) = \{n \in \mathbb{Z} | \exists k \in \mathbb{Z}, n - 0 = 3k\} = \{3k | k \in \mathbb{Z}\}$
$Cl(1) = \{n \in \mathbb{Z} | \exists k \in \mathbb{Z}, n - 1 = 3k\} = \{3k + 1 | k \in \mathbb{Z}\}$
$Cl(2) = \{n \in \mathbb{Z} | \exists k \in \mathbb{Z}, n - 2 = 3k\} = \{3k + 2 | k \in \mathbb{Z}\}$

These are the only equivalence classes. Indeed, for $n \in \mathbb{Z}$, $n$ is always related to 0, 1, or 2. (It suffices to compute the remainder of $n$ by 3 to convince oneself.) For example, $89 = 3 \times 29 + 2$ so $89 \sim 2$. Thus, by what we saw previously, we will always have $Cl(n) = Cl(0)$ or $Cl(n) = Cl(1)$ or $Cl(n) = Cl(2)$. In the example, $Cl(89) = Cl(2)$.

We note that the three sets of $\mathbb{Z}/\sim$ are disjoint, which means they have no elements in common. This is very general. Suppose we have $Cl(x)$ and $Cl(y)$ two distinct classes of $E/\sim$ and $z$ belonging to both these classes. Then, $z \sim x$ and $z \sim y$, wherefrom $x \sim y$, but then $Cl(x) = Cl(y)$, which is a contradiction to our assumption that $Cl(x)$ and $Cl(y)$ are distinct.

Furthermore, we always have

$$E = \bigcup_{\Omega \in E/\sim} \Omega$$

which reads as: for all $x \in E$, there is $\Omega \in E/\sim$ such that $x \in \Omega$ (and all $\Omega$ are subsets of $E$). Indeed, for $x \in E$, $Cl(x) \in E/\sim$ and $x \in Cl(x)$.

Let us recap. All elements of $E/\sim$ are nonempty, they are pairwise disjoint, and their union is equal to all of $E$. We say that $E/\sim$ is a partition of $E$. Conversely, if we are given a partition $\mathcal{P}$ of $E$, that is a family of pairwise disjoint nonempty parts of $E$ whose union is equal to $E$, there is an equivalence relation $\sim$ such that $\mathcal{P} = E/\sim$. This relation is very simple to construct: we say that $x \sim y$ when there is an $\Omega \in \mathcal{P}$ such that $x \in \Omega$ and $y \in \Omega$.

When we use equivalence relations, we often employ the term *representative.* A representative of a class $\Omega \in E/\sim$ is simply an element of this class. We call $x \in \Omega$ a representative since it allows us to reconstruct $\Omega$ in the sense that $\Omega = Cl(x)$. Indeed, by definition of $E/\sim$, $\Omega$ can be written as $Cl(x_0)$ with $x_0 \in E$; if $x \in \Omega$, $x \sim x_0$ and thus $Cl(x) = Cl(x_0) = \Omega$. To reconstruct all of $E/\sim$, we would like to define $V = \{x_\Omega | \Omega \in E/\sim\}$, where $x_\Omega \in \Omega$ ($x_\Omega$ is a representative of $\Omega$). In the case of the example, it is easy; we can take $V = \{0, 1, 2\}$. However, when $E/\sim$ is infinite (which already assumes that $E$ is infinite), considering the $x_\Omega$ constituting $V$ is nontrivial. There are infinitely many classes, and each must have a representative. How should one choose a representative in each class? It is impossible to prove that constructing $V$ is possible, but since we cannot show that it is impossible either (constructing $V$ doesn't violate any axioms), we can assume as an axiom that it is possible to construct $V$. This is the axiom of choice. It can seem, at first glance, to be common sense but it nevertheless leads to very surprising results, the most famous of which is the Banach-Tarski paradox: "It is possible to cut up a sphere into a finite number of pieces and to reassemble these pieces into two spheres identical to the first one." Riddles 96, 97, and 98 are also surprising uses of the axiom of choice.

## Riddles Using Equivalence Relations

The concept of equivalence relation is used in riddles 96, 97, and 98. In each of these riddles, one has to introduce an equivalence relation and to consider a family of representatives (necessitating the use of the axiom of choice).

# 4 Bijections

## An Introduction to the Concept of Bijection

Let us imagine the following situation: a teacher wanting to know if there are enough chairs in his classroom asks all his students to sit down. If all students are sitting and all chairs are occupied, then there must be as many chairs as students. In this case, we say that the set of students is in a bijection with the set of chairs. Indeed, when the teacher made his students sit, he mapped each student to a chair and each chair to a student; hence, he placed the set of students and the set of chairs in a bijection.

In an informal manner, we can say that two sets $E$ and $F$ are in a bijection if it is possible to sort the elements of $E$ and $F$ into pairs with the help of a function $f : E \to F$, which we will call a bijection.

## Surjectivity, Injectivity

Let us note $E$ as the set of students and $F$ as the set of chairs. Let $f$ be a function from $E$ to $F$. For $f$ to be a bijection, we need the following:

- No two students can be on the same chair; this is the injective nature of $f$.
- No chair can be empty; this is the surjective nature of $f$.

Mathematically, injectivity is written as:

$$\forall (x_1, x_2) \in E^2, x_1 \neq x_2 \Rightarrow f(x_1) \neq f(x_2)$$

which reads as for all pairs $(x_1, x_2)$ of elements of $E$, if $x_1$ is different from $x_2$, then $f(x_1)$ is different from $f(x_2)$.

Mathematically, surjectivity is written as:

$$\forall y \in F, \exists x \in E, f(x) = y$$

which reads as for any element $y$ of $F$, there is at least one element $x$ of $E$ such that $f(x) = y$.

An arbitrary function $f : E \to F$ is a bijection if it is injective and surjective.

Consider, for example, the function $f$ going from $\{1, 2, 3\}$ into $\{4, 5, 6, 7\}$ and defined by

$$f(1) = 4, f(2) = 5, f(3) = 7$$

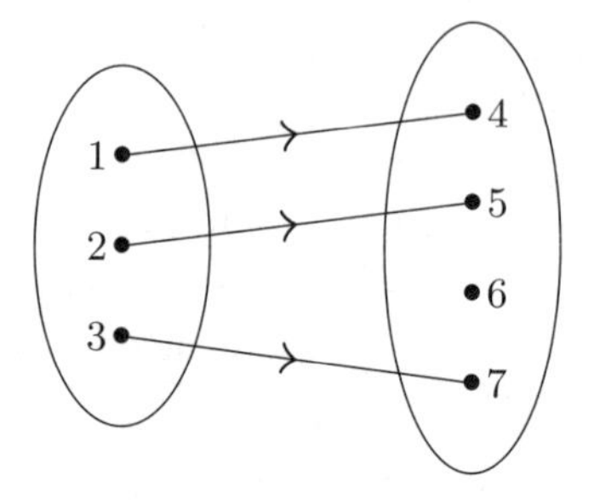


This function is injective since all the elements of its starting set $\{1, 2, 3\}$ map to distinct values. On the other hand, this function is not surjective since it doesn't map onto the whole destination set $\{4, 5, 6, 7\}$. Indeed, it never attains 6.

Consider now the function $f$ going from $\{1, 2, 3\}$ into $\{4, 5\}$ and defined by:

$$f(1) = 4, f(2) = 4, f(3) = 5$$

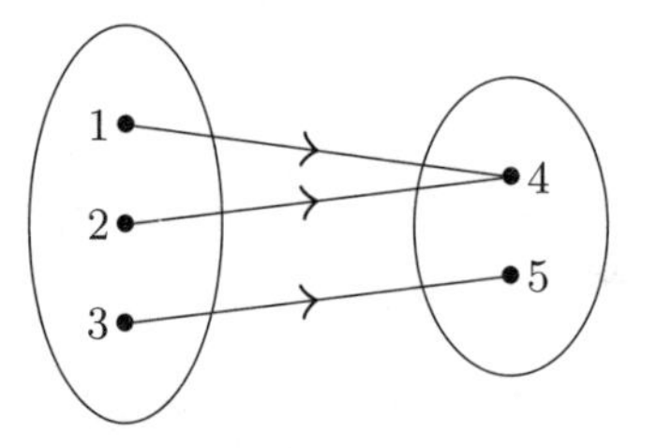


This function is not injective since 1 and 2 map to the same result. On the other hand, it is surjective since it attains all the elements of the destination set $\{4, 5\}$.

Finally, let us consider the function $f$ going from $\{1, 2, 3\}$ into $\{4, 5, 6\}$ and defined by:

$$f(1) = 4, f(2) = 5, f(3) = 6$$

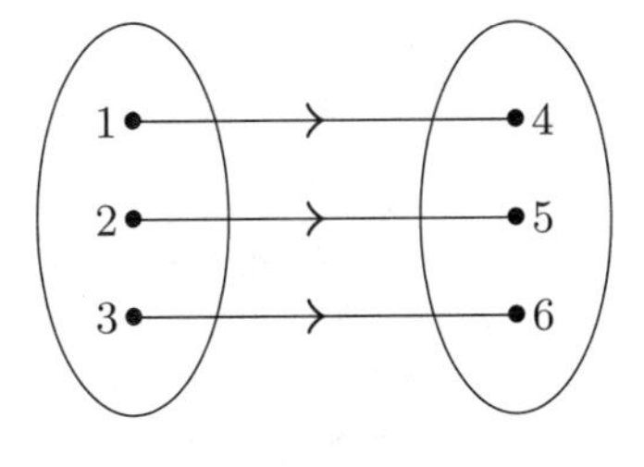


This function is both injective and surjective, so it is bijective. It sorts the elements of $\{1, 2, 3\}$ into pairs with those of $\{4, 5, 6\}$

# Examples of Bijections

Let us denote $\mathbb{N} = \{0, 1, 2, 3, \dots\}$ as the set of all integers and $\mathbb{P} = \{0, 2, 4, 6, \dots\}$ as the set of all even integers. Let us show that it is possible to put these two sets into bijection.

We can, for example, consider the following function:

$$f : \begin{cases} \mathbb{N} & \to \mathbb{P} \\ n & \mapsto 2n \end{cases}$$

In this case, $f$ is injective since for $n_1 \neq n_2$, we have $2n_1 \neq 2n_2$. Thus, $f(n_1) \neq f(n_2)$. Moreover, $f$ is surjective since all even integers are of the form $2k$ with $k \in \mathbb{N}$; hence, is attained by $f$ with $f(k) = 2k$.

Here is a more difficult example. Our goal is to construct a bijection between $\mathbb{N}$ and $\mathbb{Z} \times \mathbb{Z}$. ($\mathbb{Z} \times \mathbb{Z}$ is the set of pairs $(m, n)$ where $m$ and $n$ are whole numbers.)

The idea is to traverse $\mathbb{Z} \times \mathbb{Z}$ starting at $(0, 0)$ and spiraling outward as depicted on the figure below.

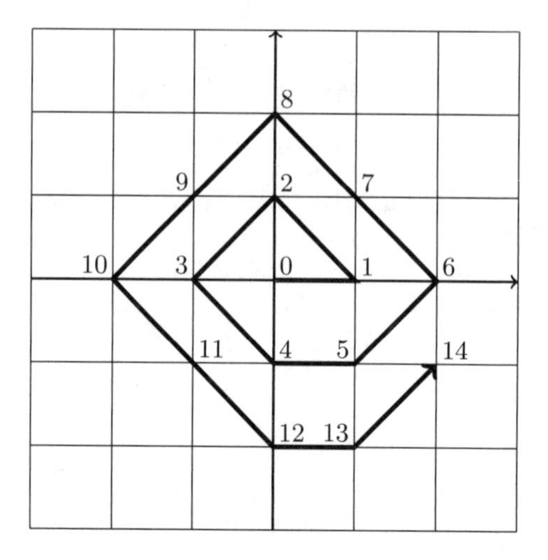


In this way, we construct a function that is both injective and surjective and thus maps each integer of $\mathbb{N}$ to a distinct element of $\mathbb{Z} \times \mathbb{Z}$. We say that $\mathbb{Z} \times \mathbb{Z}$ is countably infinite.

In general, we say that a set is countably infinite if it is in a bijection with $\mathbb{N}$. Thus, every countably infinite set has in some sense "the same size" as $\mathbb{N}$. Since $\mathbb{Z}$ and even $\mathbb{Z} \times \mathbb{Z}$ are countably infinite, one might think that all infinite sets are countably infinite. This is not the case, since the set of real numbers, noted as $\mathbb{R}$, is not countably infinite! $\mathbb{R}$ is in some sense "much bigger" than $\mathbb{N}$. In contrast, $\mathbb{R}$ is the same size as $\mathbb{R} \times \mathbb{R}$, which is pretty counterintuitive (one could think that the 2D plane has many more points than the real number line). This discovery, which had a large effect on his contemporaries, is owed to German mathematician Georg Cantor.

# Permutations

Let us return now to bijections between finite sets. Such bijections, called permutations, correspond simply to the different ways to shuffle a deck of cards.

For a nonzero integer $n$, a permutation of $\{1, \ldots, n\}$ is a bijection from $\{1, \ldots, n\}$ into $\{1, \ldots, n\}$. For example, if one takes $n = 3$, the function $\sigma$ defined on $\{1, 2, 3\}$ by $\sigma(1) = 3$, $\sigma(2) = 1$, and $\sigma(3) = 2$ is a permutation of

$\{1, 2, 3\}$. To give a permutation of $\{1, 2, 3\}$ is to give a treble $(\sigma(1), \sigma(2), \sigma(3))$ which is a shuffle of $(1, 2, 3)$. Let us list all possible trebles:

1. $(1, 2, 3)$
2. $(2, 1, 3)$
3. $(3, 2, 1)$
4. $(1, 3, 2)$
5. $(2, 3, 1)$
6. $(3, 1, 2)$

The permutation given as an example previously is permutation number 6. One can see very concretely the permutations of $\{1, 2, 3\}$ as the different ways to shuffle a deck of three cards: at the start the three cards, numbered 1 to 3, are in the order $(1, 2, 3)$, after the shuffle they are in a different order. The first permutation is the identity permutation, which involves not shuffling the deck. On the other hand, the second permutation is a real shuffle, since cards 1 and 2 have been swapped. Permutations 3 and 4 also consist in swapping two cards (respectively cards 1, 3 and cards 2, 3). Such permutations are called transpositions.

Permutation 5 is not a transposition, since there are not two but three cards that have been moved. This permutation is what we call a cycle of length three. Indeed, we can represent it below:

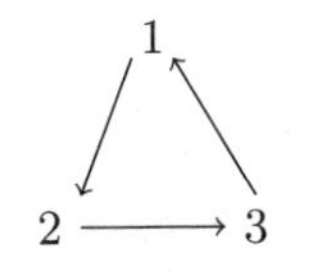


The arrows correspond to the image by the permutation. Thus 1 is sent to 2, which is sent to 3, which is sent to 1. We denote this permutation $(1\ 2\ 3)$ (without commas to differentiate it from the treble $(1, 2, 3)$). We can also denote it $(2\ 3\ 1)$, since 2 is sent to 3, which is sent to 1, which is sent to 2. Permutation 6 is another cycle of length three (or 3-cycle):

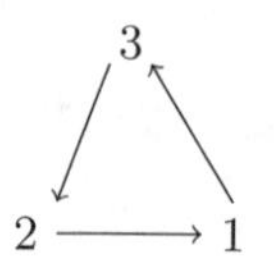


Since 3 is sent to 1, which is sent to 2, which is sent to 3, we denote this cycle (3 1 2). Do note that transpositions are cycles too, but of length two. For example, since the second permutation of the list only sends 1 to 2, which itself is sent to 1, it is denoted (1 2).

Even though it is impossible to pass from $(1, 2, 3)$ to $(2, 3, 1)$ with a single transposition, we can get there by chaining two transpositions. Indeed, if we start by applying the transposition (1 3), we go from $(1, 2, 3)$ to $(3, 2, 1)$, and, this time, by applying the transposition (2 3), we indeed arrive at $(2, 3, 1)$. The permutation applied by applying the transposition (1 3) then the transposition (2 3) is denoted $(2\ 3) \circ (1\ 3)$ ($\circ$ is the composition of maps). We thus have the following formula:

$$(1\ 2\ 3) = (2\ 3) \circ (1\ 3)$$

In fact, this is very general: one can always view a permutation of $\{1, \ldots, n\}$ as a composition of transpositions. Concretely, if we only swap cards two by two in a deck, we will still be able to make any possible shuffle. For example, we saw that (1 2 3) could be written as the composition of two transpositions; that being said, it can also be written as a composition of four transpositions, since:

$$(1\ 2\ 3) = (1\ 2) \circ (1\ 2) \circ (2\ 3) \circ (1\ 3)$$

Indeed $(1\ 2) \circ (1\ 2)$ doesn't change anything since it consists of swapping cards 1 and 2 and then swapping them right back. (1 2 3) can also be written with 6, 8, 10 transpositions. On the other hand, it will never be possible to write (1 2 3) as the composition of an odd number of transpositions. More generally, if a permutation $\sigma$ of $\{1, \ldots, n\}$ can be written as the composition of an even (respectively, odd) number of transpositions, then it will never be possible to write $\sigma$ as a composition of an odd (respectively, even) number of transpositions. In other words, the number of transpositions to obtain $\sigma$ is not unique, but its parity is. Concretely, if one shuffles a deck of cards by applying an even number of transpositions, it will be impossible to make the same shuffle with an odd number of transpositions. We omit the proof for these results since it is somewhat technical.

This allows for the definition of the sign map, used very often as soon as one deals with permutations. This map, usually denoted sgn, maps any permutation to either 1 if it can be written as the composition of an even number of permutations, or $-1$ if not (i.e., if it can be written as the composition of an odd number of transpositions). Thanks to the previously stated result, this map is well defined. For example $\operatorname{sgn}((1\ 2)) = -1$ and $\operatorname{sgn}((1\ 2\ 3)) = 1$. More generally, the motivated reader will be able to show that the signature of an $n$ cycle is $(-1)^{n-1}$.

Writing a permutation as a composition of transpositions is often useful, but it has the downside that it is not unique. One practical decomposition, which has the upside of being unique, is the decomposition into cycles of disjoint orbits. What does this mean? Let us look at the permutation $\sigma$, which to $1, \ldots, 6$, maps respectively $3, 4, 1, 5, 6, 2$. We have $\sigma = (1\ 3) \circ (2\ 4\ 5\ 6)$. Indeed, the image of $1, \ldots, 6$ by the cycle $(2\ 4\ 5\ 6)$ is $1, 4, 3, 5, 6, 2$; then, by applying the transposition $(1\ 3)$, we do get $3, 4, 1, 5, 6, 2$. We call the orbit of a cycle the set of elements on which it acts; thus, the orbit of $(1\ 3)$ is $\{1, 3\}$ and the orbit of $(2\ 4\ 5\ 6)$ is $\{2, 4, 5, 6\}$. These two sets are disjoint (which is to say they have no elements in common). We have thus decomposed $\sigma$ into two cycles of disjoint orbit, or, simply, two disjoint cycles. One can convince oneself that, up to the order of the factors (i.e., by not considering $(2\ 4\ 5\ 6) \circ (1\ 3)$ as a new decomposition), this decomposition is unique. In general, any permutation can be written as a composition of cycles with disjoint orbits, and this decomposition is unique (still up to the order of factors). This result can be proven by using the following equivalence relation: if $\sigma$ is a permutation of $\{1, \ldots, n\}$, we say that $p, q \in \{1, \ldots, n\}$ are equivalent when there is an integer $k$ such that $\sigma^k(p) = q$. The equivalence classes of these relations are the orbits of the cycles.

One last thing that can come in handy for the riddles is the combinatorics of permutations. As we saw previously, there are $6 = 3 \times 2 \times 1 = 3!$ permutations of $\{1, 2, 3\}$. More generally, there are $n!$ permutations of $\{1, \ldots, n\}$. Indeed, choosing a permutation $\sigma$ is the same as choosing $\sigma(1), \sigma(2), \ldots, \sigma(n)$. One has $n$ possibilities for $\sigma(1)$, once $\sigma(1)$ has been chosen, one has $n - 1$ possibilities for $\sigma(2)$ (since it must be different from $\sigma(1)$), then $n - 2$ possibilities for $\sigma(3)$, and so on until there is only one possibility for $\sigma(n)$. There are thus $n \times (n - 1) \times (n - 2) \times \cdots \times 1 = n!$ permutations.

Among these permutations, one could try counting how many are $n$-cycles. The $n$-cycles are of the form $(1\ c_1\ \ldots\ c_{n-1})$, where the $c_i$ are pairwise distinct integers between 1 and $n - 1$. To each family of $c_i$ corresponds a unique $n$-cycle and vice versa. There are thus as many $n$-cycles as families of $c_i$; that is $(n - 1)!$.

# Riddles Using Bijections

Riddle 72 requires the decomposition into cycles of disjoint orbits, riddle 74 requires the use of the sign, and riddle 77 uses the notion of bijection directly.

# 5 Induction

## An Introduction to Induction

Imagine dominoes placed one after the other in such a way that if a domino falls, it will knock over the next one. Then we see directly that if we knock over the first domino then all the dominoes will fall. We just reasoned by induction!

More formally, we reason by induction when we want to show that a property is true for all integers. The reasoning takes two steps. We show first that if the property is true for some integer, then it is true for the next integer. This step, called the induction step, corresponds to putting the dominoes into place: if one domino falls, it knocks over the next one. Second, we show that the property holds for integer 1: we knock over the first domino. This step is known as the basis or initialization. The property is true for integer 1, thus it is true for integers 2, 3, 4, and so on. All the dominoes fall, and the property is true for all integers.

As a first example, let us show by induction that the sum of the $n$ first odd numbers is equal to $n^2$. For example, with $n = 1$, this property says that 1 is equal to $1^2$, which is obvious. With $n = 2$, we have $1 + 3 = 2^2$, with $n = 3$, $1 + 3 + 5 = 3^2$, and so on. To show this property, we fix an arbitrary integer $n \geq 1$ and assume that $1 + 3 + ... + (2n - 1) = n^2$. We want to show that $1 + 3 + ... + (2n - 1) + (2n + 1) = (n + 1)^2$. It is easy, $1 + 3 + ... + (2n - 1) + (2n + 1) = n^2 + (2n + 1) = (n + 1)^2$. We saw that the property was true at rank 1 (and even at ranks 2 and 3), so it is true for all integers.

## Strong Induction

With the principle of induction shown above, we proved a property by showing the property was true for $\mathcal{P}_1$ and that for any integer $n \geq 1$, if $\mathcal{P}_n$ is true then, $\mathcal{P}_{n+1}$ is too.

Strong induction consists of "using all the fallen dominoes to knock over the next one": we show $\mathcal{P}_1$ and we show that for any $n$ if the properties $\mathcal{P}_1, \ldots, \mathcal{P}_n$ hold, then the property $\mathcal{P}_{n+1}$ does too.

In reality, it is not more general: we can reduce this to a weak (i.e., standard) induction. Indeed, it suffices to consider the property $\mathcal{P}'_n$: $[\mathcal{P}_1, \ldots, \mathcal{P}_n \text{ true}]$. Proving $\mathcal{P}_n$ with a strong induction is proving $\mathcal{P}'_n$ with a weak induction. Thus, the concept of strong induction has only an aesthetic value.

# An Example of Induction

For a nonzero integer $n$, let $H_n$ be the sum of the reciprocals of the integers less than or equal to $n$: $H_n = 1 + \frac{1}{2} + ... + \frac{1}{n}$. For example, $H_1 = 1$, $H_2 = \frac{3}{2}$, $H_3 = \frac{11}{6}$, and so on. The sequence $(H_n)_{n \geq 1}$ has an interesting mathematical property (which we will not prove here and which is necessary only to understand the motivation of the question studied): it goes to $+\infty$. Moreover, all its terms are clearly rational numbers. This naturally leads to the question: which are the integer terms? In fact, only $H_1$ is an integer.

We will in fact show that for any integer $n \geq 2$, $H_n$ is not an integer.

Let $\mathcal{P}_n$ "$H_n = \dfrac{i_n}{p_n}$, with $i_n$ odd and $p_n$ even." It suffices to show that $\mathcal{P}_n$ holds: since $i_n$ is not divisible by 2, if $H_n = \dfrac{i_n}{p_n}$, $H_n$ is not an integer.

First we show $\mathcal{P}_2$: $1 + \frac{1}{2} = \frac{3}{2}$.

Let us now assume $\mathcal{P}_2, \ldots, \mathcal{P}_n$. Let us show $\mathcal{P}_{n+1}$. We must separate two cases:

- If $n+1$ is odd, $H_{n+1} = H_n + \frac{1}{n+1} = \dfrac{i_n(n+1) + p_n}{p_n(n+1)}$, the numerator is odd and the denominator is odd; hence, $\mathcal{P}_{n+1}$.

- If $n+1$ is even, $n+1 = 2m$ with $m$ an integer greater than or equal to 2. $H_{n+1} = \frac{1}{2}H_m + 1 + \frac{1}{3} + ... + \frac{1}{n} = \frac{1}{2}H_m + \frac{k}{l}$, with $l = 1 \times 3 \times ... \times n$, thus $l$ is odd, and $k$ is an integer. $H_{n+1} = \dfrac{i_m l + 2kp_m}{2p_m l}$, hence $\mathcal{P}_{n+1}$.

Finally, by induction we have the announced result.

## An Example of Fallacious Induction

Reasoning by induction hides some subtleties one must know. Let us see an example of an erroneous reasoning by induction.

Let us show by induction that $n$ points in the 2D plane are always aligned.

Basis: For $n = 1$, a point in the plane is always aligned. We can even start with $n = 2$: two points in the plane are always aligned.

Induction: Let $n$ be an arbitrary integer; we assume $n$ points in the plane are always aligned. For this, we consider a group of $n$ points among the $n + 1$ points in the plane; by the induction hypothesis, these $n$ points are aligned. Now we consider another group of $n$ points in the plane that are also aligned. The two lines thus obtained must be the same, so finally the $n + 1$ points are all aligned. Hence the result...which is absurd!

The basis is valid and so is the induction. However, the induction is only valid for $n \geq 3$ when the basis is only valid for $n = 1$ and $n = 2$. In other words, we knocked over the first two dominoes and showed that from the third one on, if a domino falls it knocks over the next one. This is not sufficient to knock over all the dominoes, since no domino knocked over the third one.

## Riddles Using Induction

Induction reasoning is omnipresent in mathematics. When a statement contains an arbitrary $n$, one can always try to do an induction on $n$. In some statements, $n$ is fixed and thus "hidden"; for example, if the question pertains to "100 students," we can still take $n = 100$ and do an induction on $n$. Riddles 24, 30, and 44 are classic examples of induction. Riddles 35, 66, 83, and 92 also use an induction.

# 6 Congruence

## The Concept of Divisibility

Before talking about congruence, we must speak of divisibility. When we divide 8 by 2, we find that $\frac{8}{2} = 4$, which is a whole number. Thus 2 *divides* 8. On the other hand, if we divide 8 by 3, we find $\frac{8}{3} \simeq 2.67$. The result is not a whole number, so we say that 3 does not divide 8. Likewise, 2 divides 26, 2 divides $-144$, but 2 does not divide 77.

Formally, we say a whole number $a$ divides a whole number $b$ if there is a whole number $k$ such that $b = k \times a$. To return to the first example, 2 divides 8 since 8 can be written as $4 \times 2$. On the other hand, 3 does not divide 8 since there is no whole number $k$ such that $8 = k \times 3$. Do note that the negative whole numbers can be considered: 2 divides $-144$ since there is a whole number $k = -72$ such that $-144 = k \times 2$.

Note that saying that $a$ divides $b$ is the same as saying that $b$ is a multiple of $a$. Indeed, we call $b$ a multiple of $a$ if there is a whole number $k$ such that $b = k \times a$. For example, 26 and $-144$ are multiples of 2.

If a whole number $n \geq 2$ is such that only 1 and $n$ divide $n$, $n$ is called a prime number. The first prime numbers are $2, 3, 5, 7, 11, ....$ If $a$, $b$ are two whole numbers, if 1 is the only common divider, we say that $a$ and $b$ are coprime. For example, $9, 16$ are coprime, but $9, 6$ are not (3 is a common divider).

## Definition of Congruence

Congruence is a very important tool in arithmetic because it can simplify many problems of divisibility. One can see congruence as weakened equality. The whole numbers 23 and 8 are not equal, but they are linked by the fact that their difference $23 - 8$ (or $8 - 23$) is a multiple of 5. We say that 23 and 8 are *congruent modulo* 5, and we also say that 23 *is congruent to* 8 modulo 5 or that 8 *is congruent to* 23 modulo 5. On the other hand, 23 and 8 are not *congruent modulo* 6 since their difference is not a multiple of 6.

More generally, if we consider a whole number $n$ ($n = 5$ in the previous example), we say that *$a$ and $b$ are congruent modulo $n$* when there is a whole number $k$ such that $a = b + kn$. This translates the fact that the difference $a - b$ is a multiple of $n$. We then note $a \equiv b\,[n]$. It is thus no different to say that $a \equiv b\,[n]$, that $a - b$ is a multiple of $n$, or that $n$ divides $a - b$.

For example, $13 \equiv 1\,[2]$ since $13 - 1$ is a multiple of 2, or, in accordance with the definition, $13 = 1 + 6 \times 2$. In this manner, we have $111 \equiv 1\,[2]$, $10 \equiv 1\,[3]$, $61 \equiv 12\,[7]$ but $13 \not\equiv 2\,[9]$ and $24 \not\equiv 5\,[13]$.

## Properties of the *congruent with* Relation

Let us now establish a short list of the properties of the relation *congruent with*, which we will show each time by returning to the formal definition of $a$ and $b$ congruent modulo $n$ when there is a whole number $k$ such that $a = b + kn$.

1. For any whole number $a$, $a \equiv a\,[n]$.
2. If $a \equiv b\,[n]$, $b \equiv a\,[n]$.
3. If $a \equiv b\,[n]$ and $b \equiv c\,[n]$, then $a \equiv c\,[n]$.
4. If $a \equiv b\,[n]$ and $c \equiv d\,[n]$, then $a + c \equiv b + d\,[n]$. We say that congruence is compatible with addition.
5. If $a \equiv b\,[n]$ and $c \equiv d\,[n]$, then $ac \equiv bd\,[n]$. We say that congruence is compatible with multiplication.
6. For any whole number $m$, if $a \equiv b\,[n]$, then $ma \equiv mb\,[n]$.
7. For any whole number $q$, if $a \equiv b\,[n]$, then $a^q \equiv b^q\,[n]$.

**Proof**

1. $a = a + 0 \times n$.

2. Suppose $a \equiv b\,[n]$. There is a whole number $k$ such that $a = b + kn$. $b = a + (-k)n$, thus $b \equiv a\,[n]$.

3. There are $k$ and $k'$, two whole numbers such that $a = b + kn$ and $b = c + k'n$; hence, $a = c + (k + k')\,n$.

4. We have $k$ and $k'$, two whole numbers such that $a = b + kn$ and $c = d + k'n$; hence, $a + c = b + d + (k + k')\,n$.

5. $ac = bd + (dk + bk' + kk'n)\,n$.

6. Consequence of the compatibility with multiplication ($m \equiv m\,[n]$).

7. We can show this by induction on $q \geq 0$. The basis is clear (for $q = 0$, $a^q = b^q = 1$), and induction follows from the compatibility with multiplication: if $a^q \equiv b^q\,[n]$, as $a \equiv b\,[n]$, $a^q a \equiv b^q b\,[n]$, which is $a^{q+1} \equiv b^{q+1}\,[n]$.

Properties 4, 5, 6, and 7 are only implications: their converses are false. Indeed, $3 \times 1 \equiv 3 \times 2\,[3]$ and $1^2 \equiv 2^2\,[3]$ but $1 \not\equiv 2\,[3]$.

Let us now mention (without proof) the famous Bézout's identity: if $a$ and $b$ are coprime, then there exist $c$, $d$ such that $ab + cd = 1$. Reciprocally, if there exist $c$, $d$ such that $ab + cd = 1$, then $a$, $b$ are coprime.

## An Application to Cryptography

Modular arithmetic has numerous applications in cryptography. The simplest is without doubt the Caesar Cypher. The concept is to replace each letter by the letter three steps further in the alphabet. Thus, A becomes D, B becomes E, C becomes F, and so on. To loop past the end, X becomes A, Y becomes B, and Z becomes C.

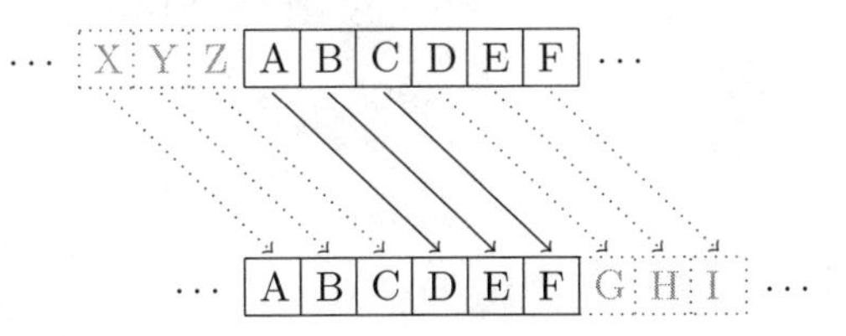


With the help of modular arithmetic, we can write the encryption/decryption algorithm explicitly. We begin by mapping each letter to its position in the alphabet (a number between 0 and 25).

- Encoding: to $i \in \{0, ..., 25\}$ map $i+3$ modulo 26
- Decoding: to $i \in \{0, ..., 25\}$ map $i-3$ modulo 26

## An Application to a Divisibility Problem

The Fermat numbers are numbers of the form $2^{2^n}+1$ where $n$ is an whole number. They are named after French mathematician Pierre de Fermat, who conjectured that they were all prime. Indeed, $F_0 = 3$, $F_1 = 5$, $F_2 = 17$, $F_3 = 257$, and $F_4 = 65537$ are in fact prime. The Fermat numbers grow very quickly, making the question of their primality very delicate. Nevertheless, Euler showed that 641 divides $2^{32}+1$ and thus that $F_5$ is not prime, without even computing $2^{32}+1$; he simply used the properties of the relation *congruent with.*

To do this, he wrote: $641 = 640+1 = 5 \times 2^7 + 1$ and thus $5 \times 2^7 \equiv -1[641]$ (by definition), and thus $5^4 \times 2^{28} \equiv 1[641]$ (property 7 with $q = 4$).

But, $641 = 16 + 625 = 2^4 + 5^4$ so $5^4 \equiv -2^4[641]$; thus, by property 6, $5^4 \times 2^{28} \equiv -2^4 \times 2^{28}[641]$. From the above, $1 \equiv -2^4 \times 2^{28}[641]$ (property 3) so by adding $2^{32}$ on either side (property 4), $2^{32}+1 \equiv 0\,[641]$, which means 641 divides $2^{32}+1$, the claimed result.

Today, we think (but it has not been proven) that the only Fermat numbers that are prime are $F_0$, $F_1$, $F_2$, $F_3$, and $F_4$ !

## An Applications to Infeasible Computations

The cited properties of congruence can significantly simplify some computations. For example, we can easily find the last digit of the decimal form of $343^{1000}$. We notice that this number is congruent to $343^{1000}$ modulo 10 (see the reminder on bases). Then, since $343 \equiv 3\,[10]$, $343^{1000} \equiv 3^{1000}\,[10]$ (property 7). $3^2 \equiv -1\,[10]$, so $3^{2\times 500} \equiv (-1)^{500}\,[10]$: $3^{1000} \equiv 1\,[10]$. In other words, the last digit of $343^{1000}$ is 1.

## Riddles Using Congruence

Riddle 55 uses the concept of divisibility, whereas riddles 39, 73, 75, and 80 use congruence.

# 7 Bases of Number Systems

## Base 10

We count in base 10. The base 10 number system is a way of representing numbers by using 10 digits: $0, 1, 2, 3, 4, 5, 6, 7, 8, 9$. For example, $346 = 3 \times 100 + 4 \times 10 + 6 = 3 \times 10^2 + 4 \times 10^1 + 6 \times 10^0$. More generally if $c_0, \ldots, c_n$ are digits,

$$c_n c_{n-1} \ldots c_1 c_0 = \sum_{k=0}^{n} c_k \times 10^k$$

## Base 2

Base 2 only uses two digits: 0 and 1. Otherwise, it works the same as base 10. To avoid confusion, all base 2 numbers will be denoted by an index $_2$. For example, $10011_2$ represents in base 2 the number:

$$\begin{aligned} 10011_2 &= 1 \times 2^4 + 0 \times 2^3 + 0 \times 2^2 + 1 \times 2^1 + 1 \times 2^0 \\ &= 16 + 2 + 1 \\ &= 19 \end{aligned}$$

When we count in base 2, we obtain the following sequence:

$$\begin{array}{ccccc} 0_2 & = & 0 \times 2^0 & = & 0 \\ 1_2 & = & 1 \times 2^0 & = & 1 \\ 10_2 & = & 1 \times 2^1 + 0 \times 2^0 & = & 2 \\ 11_2 & = & 1 \times 2^1 + 1 \times 2^0 & = & 3 \\ \vdots & & \vdots & & \vdots \\ 1010_2 & = & 1 \times 2^3 + 0 \times 2^2 + 1 \times 2^1 + 0 \times 2^0 & = & 10 \end{array}$$

Like in base 10, if $c_0, \ldots, c_n$ are digits (i.e., zeroes and ones), then

$$(c_n c_{n-1} \ldots c_1 c_0)_2 = \sum_{k=0}^{n} c_k \times 2^k$$

# Numbering in Base 2

Many combinatorial problems can be simplified by using base 2. The idea is often the same: number the objects of the problem in base 2. Let us see how this method can lead to a solution on a concrete example.

Let us consider the following problem. We have 16 sheets of paper and four colored pens, which are yellow, red, blue, and green. The goal is to draw a yellow line on eight sheets, a red line on eight sheets, a blue line on eight sheets, and a green line on eight sheets, while ending up with all possible configurations, that is one sheet with no line (not drawn upon), one sheet with only a yellow line, ..., one sheet with three lines (e.g., yellow, green, and blue), and so on.

The simplest solution is to number the 16 sheets in base 2. The first sheet is sheet number $0000_2$, followed by sheets $0001_2$, $0010_2$, $0011_2$, $0100_2$, and so on, all the way to sheet $1111_2$. Now it suffices to mark with the yellow pen all the sheets whose first digit (starting from the right) is a 1: sheet numbers $0001_2$, $0011_2$, $0101_2$, $0111_2$, $1001_2$, $1011_2$, $1101_2$, $1111_2$. Likewise we mark with the red pen all the sheets whose second digit is a 1: sheet numbers $0010_2$, $0011_2$, $0110_2$, $0111_2$, $1010_2$, $1011_2$, $1110_2$, $1111_2$. We mark in blue the sheets numbered $0100_2$, $0101_2$, $0110_2$, $0111_2$, $1100_2$, $1101_2$, $1110_2$, $1111_2$. Finally, we mark in green the sheets numbered $1000_2$, $1001_2$, $1010_2$, $1011_2$, $1100_2$, $1101_2$, $1110_2$, $1111_2$. In this way, we notice that in the end, the sheet number $(c_3c_2c_1c_0)_2$ is marked in yellow if and only if $c_0 = 1$, is marked in red if and only if $c_1 = 1$, and so on. We can thus conclude that all possible combinations appear; for example, the unmarked sheet is sheet $0000_2$, the sheet marked with only yellow is sheet number $0001_2$, and the sheet marked with all colors is the sheet $1111_2$. The problem is solved.

# Riddles Using Base 2 Numbering

Riddles 46, 57, 83, and 95 use numbering in base 2.

# 8 Some Probabilities

## Independence and Conditional Probabilities

Two events are independent when they have no influence on one another. For example, if we are playing heads or tails, the result of the first flip has no influence on the result of the second flip. Likewise, with a die, obtaining a result greater than or equal to 5 doesn't influence the probability of obtaining an even result (which remains equal to $\frac{1}{2}$). On the other hand, obtaining a result greater than or equal to 4 influences the probability that the result is even (which becomes $\frac{2}{3}$).

Let us now move to conditional probabilities. The probability that event A happens given that event $B$ happens is denoted $\mathbb{P}(\mathrm{A}|\mathrm{B})$ and reads "probability of A given B." It is a conditional probability. For example, if we denote A as the event "it snows today" and B as the event "it snowed yesterday," it is clear that $\mathbb{P}(\mathrm{A}|\mathrm{B}) > \mathbb{P}(\mathrm{A})$: if it snowed the previous day, there is a greater chance that it will snow again today (since there is a higher probability that we are in winter). On the other hand, if we play heads or tails, denoting A as "the next flip is tails" and B as "the previous flip was heads," we have $\mathbb{P}(\mathrm{A}|\mathrm{B}) = \mathbb{P}(\mathrm{A})$, so we say that both events are independent. By definition, for two events A and B such that $\mathbb{P}(\mathrm{B}) \neq 0$, we have:

$$\mathbb{P}(\mathrm{A}|\mathrm{B}) = \frac{\mathbb{P}(\mathrm{A \text{ and } B})}{\mathbb{P}(\mathrm{B})}$$

This formula is intuitive: we constrain ourselves to set B by applying the formula:

$$\text{probability} = \frac{\text{number of favorable cases}}{\text{number of possible cases}}$$

Note that to condition on event B, we must have $\mathbb{P}(\mathrm{B}) \neq 0$; if $\mathbb{P}(\mathrm{B}) = 0$, then assuming event B has happened is meaningless.

In summary, by definition, two events A and B are independent if $\mathbb{P}(A|B) = \mathbb{P}(A)$, which is equivalent to $\mathbb{P}(A \text{ and } B) = \mathbb{P}(A) \times \mathbb{P}(B)$.

## Bayes' Rule

Bayes' rule allows us to relate $\mathbb{P}(A|B)$ and $\mathbb{P}(B|A)$ for two events A and B of nonzero probability

$$\mathbb{P}(A \text{ and } B) = \mathbb{P}(A|B) \times \mathbb{P}(B)$$

but

$$\mathbb{P}(A \text{ and } B) = \mathbb{P}(B \text{ and } A) = \mathbb{P}(B|A) \times \mathbb{P}(A)$$

hence

$$\mathbb{P}(A|B) = \frac{\mathbb{P}(B|A) \times \mathbb{P}(A)}{\mathbb{P}(B)}$$

This is Bayes' rule.

Let us see how to apply Bayes' rule in a concrete case. A box contains two coins, one fair coin with a head (H) and a tail (T), and one unfair coin with two H. We choose a coin at random and place it on the table. The visible side is H. What is the probability that the other side is T?

Very often, the answer given is "50/50," with the reason that the coin is either the fair one or the unfair one; since we don't know which it is, the second face could be either H or T. Of course, this reasoning is fallacious. Note T as the event "the hidden side of the picked coin is T" and H as "the visible side of the picked coin is H." Let us apply Bayes' rule:

$$\mathbb{P}(T|H) = \frac{\mathbb{P}(H|T) \times \mathbb{P}(T)}{\mathbb{P}(H)}$$

$\mathbb{P}(H|T) = 1$ if the hidden side is is T; then the chosen coin is the fair coin and the visible side is necessarily H.

$\mathbb{P}(T) = \frac{1}{4}$ the hidden face of the coin being chosen at random among four possible faces, three H and one T.

$\mathbb{P}(H) = \frac{3}{4}$ the visible side of the coin being chosen at random among four possible faces, three H and one T.

In the end, the sought after probability is $\mathbb{P}(T|H) = \frac{1}{3}$.

This proof is not the best way to convince oneself of the result. The simplest is to notice that when a coin is drawn at random and placed on the table, there are four equiprobable cases:

1. The coin chosen is fair, and the visible face is T.
2. The coin chosen is fair, and the visible side is H.
3. The coin chosen is unfair, and the visible face is its first H face.
4. The coin chosen is unfair, and the visible face is its second H face.

Knowing that the visible face is H, we are in case 2, 3, or 4. The coin has a hidden face T only in case 2, so in 1-in-3 cases.

There is nonetheless a point to clarify: why do cases 2, 3, and 4 remain equiprobable? It is because we condition relative to the event "the visible face is H," which is the union of three cases 2, 3, and 4. More generally, if we have a partition of the sample space $\Omega$ into $n$ equiprobable cases $A_1, \ldots, A_n$ and $I \subset \{1, \ldots, n\}$, for $k \in I$, the probabilities $\mathbb{P}\left(A_k | \bigcup_{i \in I} A_i\right)$ are all equal. (The union is equivalent to "or"; the intersection to "and.")

$$\mathbb{P}\left(A_k | \bigcup_{i \in I} A_i\right) = \frac{\mathbb{P}\left(A_k \bigcap \left(\bigcup_{i \in I} A_i\right)\right)}{\mathbb{P}\left(\bigcup_{i \in I} A_i\right)} = \frac{\mathbb{P}\left(A_k\right)}{\mathbb{P}\left(\bigcup_{i \in I} A_i\right)}$$

This isn't so obvious since if we condition relative to an event that spans more than one $A_i$, then it need not be true anymore.

# Riddles Using Probability

In this book, a high number of riddles are related to probability. In particular, riddles 25, 31, 32, 59, 85, and 86 use conditional probability and Bayes' rule. Furthermore, riddles 31, 32, 50, and 86 explain false reasonings on probabilities that one must avoid.

# Bibliography

BOLLOBAS B. *The Art of Mathematics: Coffee Time in Memphis.* Cambridge University Press, 2006.

CHERNYAK Y. *The Chicken from Minsk: And 99 Other Infuriating Brainteasers.* Basic Books, 1997.

ENGEL A. *Problem-Solving Strategies.* Springer, 1998.

GARDNER M. *My Best Mathematical and Logic Puzzles.* Dover Publications,1994.

KORDEMSKY B. *The Moscow Puzzles: 359 Mathematical Recreations.* Dover Publications, 1992.

WINKLER P. *Mathematical Puzzles: A Connoisseur's Collection.* AK Peters, 2003.

"Mathproblems.info," Michael Shackleford, accessed August 18, 2022, http://mathproblems.info

"Nick's Mathematical Puzzles," Nick Hobson, last updated February 26, 2009, http://www.qbyte.org/puzzles/

"Un peu de probabilités," David Madore, last updated July 2, 2008, http://www.madore.org/~david/math/proba.html

# Index